SCHISTOSOMIASIS

World Class Parasites

VOLUME 10

Volumes in the World Class Parasites book series are written for researchers, students and scholars who enjoy reading about excellent research on problems of global significance. Each volume focuses on a parasite, or group of parasites, that has a major impact on human health, or agricultural productivity, and against which we have no satisfactory defense. The volumes are intended to supplement more formal texts that cover taxonomy, life cycles, morphology, vector distribution, symptoms and treatment. They integrate vector, pathogen and host biology and celebrate the diversity of approach that comprises modern parasitological research.

Series Editors

Samuel J. Black, *University of Massachusetts, Amherst, MA, U.S.A*
J. Richard Seed, *University of North Carolina, Chapel Hill, NC, U.S.A.*

SCHISTOSOMIASIS

Edited by

W. Evan Secor
Division of Parasitic Diseases, National Center for Infectious Diseases, Centers for
Disease Control and Prevention

and

Daniel G. Colley
Center for Tropical and Emerging Global Diseases and Department of Microbiology,
University of Georgia

 Springer

Cover illustration: Adult schistosome worm pair *in copulo*; egg of *Schistosoma mansoni*, demonstrating lateral spine. Drawing by Peter Augostini.

Library of Congress Cataloging-in-Publication Data

Schistosomiasis / edited by W. Evan Secor and Daniel G. Colley.
    p. cm. – (World class parasites ; v. 10)
  Includes bibliographical references and index.
  ISBN 0-387-23277-X (alk. paper)
  1. Schistosomiasis. I. Secor, W. Evan, 1961- II. Colley, Daniel G., 1943- III. Series.

RC182.S24S322 2005
616.9'63—dc22

2004058995

Printed in the United States of America.

9 8 7 6 5 4 3 2 1   SPIN 11052791

springeronline.com

Contents

Preface

Human schistosomiasis is a disease with a rich and well-documented past, and every expectation of an unfortunately long future. These infections were known to the ancient Egyptians and their transmission shows little evidence of slowing down, globally. The good news is that field applicable, and increasingly affordable, chemotherapy has been available for almost 25 years. Using chemotherapy and other means of control, some countries have decreased transmission and made excellent headway against morbidity. The bad news is that the public health problems caused by schistosomiasis are still with us, with the estimated number of cases of schistosomiasis, while shifting geographically, remaining approximately 200 million for the last 30 years. In fact, with the development of field usable ultrasound technology and meta-analyses performed on existing data, there is a new appreciation for the extent of non-lethal morbidity associated with these infections. While the percentage of individuals with severe hepatosplenic disease remains below 10%, recent reassessments of morbidity associated with schistosomiasis indicate that the prevelance of symptoms and the cost in diability-adjusted life years is much greater than was previously, commonly appreciated (Van der Werf, M. J., et al. 2003, *Acta Tropica* 86:125-139; Charles H. King, personnel communication). Strong impetus for addressing these issues is provided by the World Health Assembly's recently passed Resolution 54.19, which calls for efforts to reduce morbidity caused by schistosomiasis and soil-transmitted helminths in school-aged children, largely through chemotherapy campaigns.

World Health Assembly recognition of the public health problems caused by schistosome infections promotes efforts towards schistosomiasis morbidity control. However, for a long-term, permanent solution, an equally needed push is to encourage continued research by a community of scientists, to learn more about multiple aspects of the parasites and disease. Unfortunately, the number of students and scholars who train in schistosomiasis, as is the case for most helminthic diseases, is decreasing. For continued progress, the field needs a critical mass of researchers investigating everything from the basic biology of the parasite to the best means by which to apply suitable public health interventions. Solid research begets useful public health tools and advances in the laboratory, even if not directly applicable to field work, provide benefits towards an overall understanding of the disease and its control. In addition, many discoveries made through the study of schistosomiasis have had profound scientific impact on other fields of biomedical research.

The goal of this book is to provide the reader with insights into the active research and programs currently related to schistosomiasis, and to use these insights as a way to project forward into the next 10-15 years of work on schistosomiasis, spanning the spectrum from research to public health interventions. The charge given to each of the authors of the chapters in the book has been to think and write freely about their work, with an emphasis on how it fits into current thinking, and where it might take us in the future. Through this we hope to bring heightened focus, and raise expectations, related to schistosomiasis, especially among trainees in the world of biomedicine and public health. A secondary goal of this volume is that it will initiate conversations among those working across the research-to-control spectrum on schistosomiasis about the future of their field, and by doing so lead to constructive efforts to identify and address the most critical questions and challenges related to schistosomiasis.

In the tradition of David Rollinson and Andrew Simpson (*The Biology of Schistosomes*, Academic Press, 1988), we organized this volume in an approach that takes the reader from genes to latrines. The first 4 chapters address schistosome phylogenetics, gene expression, and the overall genome, including information on exciting new tools for addressing questions that have long been inaccessible to schistosomologists. The next 3 chapters explore the host-schistosome interaction at the larval to adult worm interface and addresses aspects important for vaccine development as well as how differential gene expression as detected by DNA microarrays may be utilized to develop tools for detection and control of infection or pathology. The following 3 chapters explore the development of the host immune response to eggs, granuloma formation and factors affecting the development and regulation of immunopathology. The next 4 chapters address the public health concerns associated with schistosomiasis, including morbidity control, host genetics, treatment and proposals for improved partnerships. The volume concludes with a chapter addressing the schisms that sometimes exist along the spectrum from basic research programs to the implementation of control schemes, and a proposal to make these differences benefit patients and researchers rather than succumb to base temptations to compete for resources to no one's benefit.

Like many of the diseases featured in the World Class Parasites series, the prospects for dramatic advances in schistosomiasis coincide with a seemingly shrinking pool of both human and material resources. Our hope is to point out that these challenges are not insurmountable, and are in fact exciting, and that the various disciplines employed in the study of schistosomiasis can and should be shaped by the needs identified by other disciplines. The most meaningful progress will occur as the laboratory better understands the needs in the field and the field better understands the capabilities of the laboratory. It is our desire that this volume contributes to the conversation in a useful, collegial manner.

W. Evan Secor and Daniel G. Colley

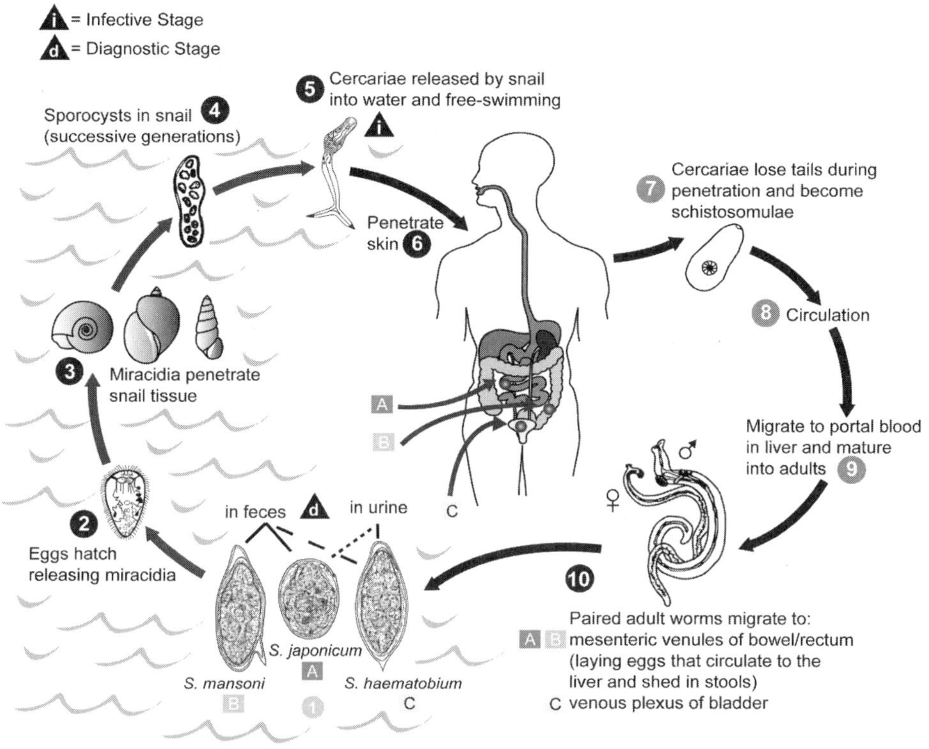

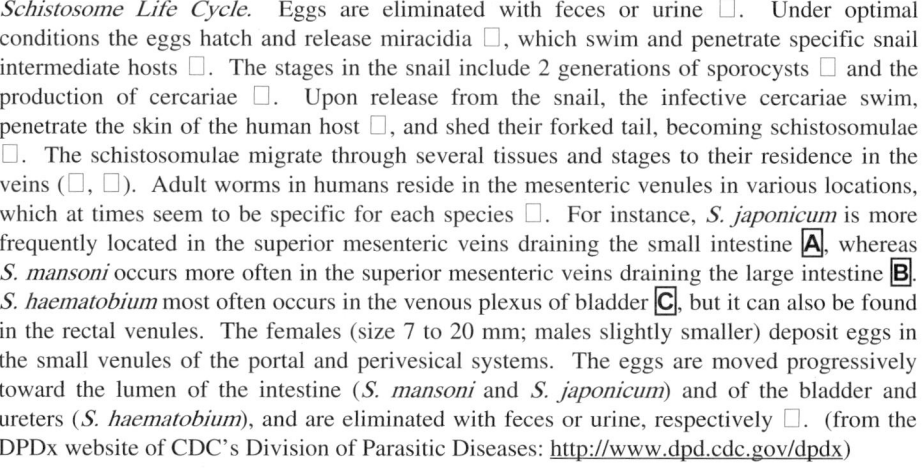

= Infective Stage

= Diagnostic Stage

Sporocysts in snail **4**
(successive generations)

Cercariae released by snail **5**
into water and free-swimming

Penetrate
skin **6**

Cercariae lose tails during
7 penetration and become
schistosomulae

8 Circulation

3 Miracidia penetrate
snail tissue

Migrate to portal blood
in liver and mature
into adults **9**

in feces   in urine   C

2
Eggs hatch
releasing miracidia

S. japonicum
A
S. mansoni   *S. haematobium*
B   C

10

Paired adult worms migrate to:
A B mesenteric venules of bowel/rectum
(laying eggs that circulate to the
liver and shed in stools)
C venous plexus of bladder

Schistosome Life Cycle. Eggs are eliminated with feces or urine □. Under optimal conditions the eggs hatch and release miracidia □, which swim and penetrate specific snail intermediate hosts □. The stages in the snail include 2 generations of sporocysts □ and the production of cercariae □. Upon release from the snail, the infective cercariae swim, penetrate the skin of the human host □, and shed their forked tail, becoming schistosomulae □. The schistosomulae migrate through several tissues and stages to their residence in the veins (□, □). Adult worms in humans reside in the mesenteric venules in various locations, which at times seem to be specific for each species □. For instance, *S. japonicum* is more frequently located in the superior mesenteric veins draining the small intestine **A**, whereas *S. mansoni* occurs more often in the superior mesenteric veins draining the large intestine **B**. *S. haematobium* most often occurs in the venous plexus of bladder **C**, but it can also be found in the rectal venules. The females (size 7 to 20 mm; males slightly smaller) deposit eggs in the small venules of the portal and perivesical systems. The eggs are moved progressively toward the lumen of the intestine (*S. mansoni* and *S. japonicum*) and of the bladder and ureters (*S. haematobium*), and are eliminated with feces or urine, respectively □. (from the DPDx website of CDC's Division of Parasitic Diseases: http://www.dpd.cdc.gov/dpdx)

Chapter 1

SCHISTOSOMES AND THEIR SNAIL HOSTS
The Present and Future of Reconstructing Their Past

Eric S. Loker[1] and Gerald M. Mkoji[2]
*[1]Department of Biology, University of New Mexico, Albuquerque, New Mexico 87131;
[2]Center for Biotechnology Research and Development, Kenya Medical Research Institute,
P.O. Box 54840, Nairobi, Kenya*

Key words: evolution, phylogenetics, host shifts, Platyhelminthes, Schistosomatidae

1. INTRODUCTION: PARASITES EXTRAORDINAIRE

The schistosomes are "World Class" parasites in every sense of the expression. They are among the most common and debilitating infectious agents of humans and their domestic animals – they still quietly infect an estimated 200 million people, especially children, and 165 million head of livestock (DeBont and Vercruysse, 1998; Chitsulo et al., 2000). The schistosomes have a geographic distribution that encompasses much of the world and their intrinsically fascinating biology poses questions of fundamental significance to both basic and applied biologists around the globe. Although many of the mysteries pertaining to the evolutionary history of this important parasite group will no doubt remain permanently shrouded in the past, there are today several new insights regarding their history that have emerged from molecular phylogenetic studies, and it is these that we wish to highlight below. We also attempt to identify particular studies that remain to be done. The fundamental premise is that by understanding their past, we can better comprehend all aspects of contemporary schistosome biology, including their likely response to ongoing environmental changes.

2.   WHERE DO SCHISTOSOMES FIT IN THE DIGENEAN FAMILY TREE?

The 13 genera and approximately 100 species of the family Schistosomatidae are parasites of crocodiles (one known species), birds and mammals. Among the 18,000 or so species of digenetic trematodes (Phylum Platyhelminthes, Class Trematoda, Subclass Digenea), the schistosomes along with the families Spirorchiidae (in turtles) and Sanguinicolidae (in fish) are odd because they have two host life cycles (a snail intermediate host and a vertebrate definitive host) that feature direct penetration of the skin of the definitive host by cercariae. They do not have the metacercaria stage and the three host life cycle that is typical of most digeneans. Representatives of these three families are also known as "blood flukes" because the adults usually live in the vascular systems of their hosts. Most species of schistosomes live in the venous system, but at least two species colonize the arterial system (McCully et al, 1967; Ulmer and Vande Vusse, 1970). Perhaps because of their vascular habitats, the tegument of adult worms of all three families is bounded by a double lipid bilayer, whereas most other flukes possess a single bilayer (McLaren and Hockley, 1977). The schistosomes also have separate sexes (are dioecious), in sharp contrast to the spirorchiids, sanguinicolids and almost all remaining digeneans that are monoecious (Platt and Brooks (1997).

Given their peculiar properties, are schistosomes modern-day descendents of digeneans that diverged early in the evolution of the group and that have retained primitive features? Or do they represent more recent offshoots that have lost primitive features and adopted peculiar features in response to life in a specialized environment? The most recent and complete phylogenetic analyses of digeneans based on SSU and partial LSU rDNA data (Fig. 1-1A) suggest that schistosomes (and other blood flukes) are indeed part of a lineage that diverged early in the evolution of digeneans, but their two-host life cycles probably are indicative of secondary loss of the third host rather than of the primitive life cycle type in digeneans (Olson et al., 2003; Cribb et al., 2003). If we assume this figure represents the true relationships among the blood flukes, then the sister group of the schistosomes are the Spirorchiidae, the turtle blood flukes (see also Blair et al., 2001), a close relationship further suggested by life cycle and morphological studies (Carmichael, 1984; Brooks et al., 1985; Combes, 1990).

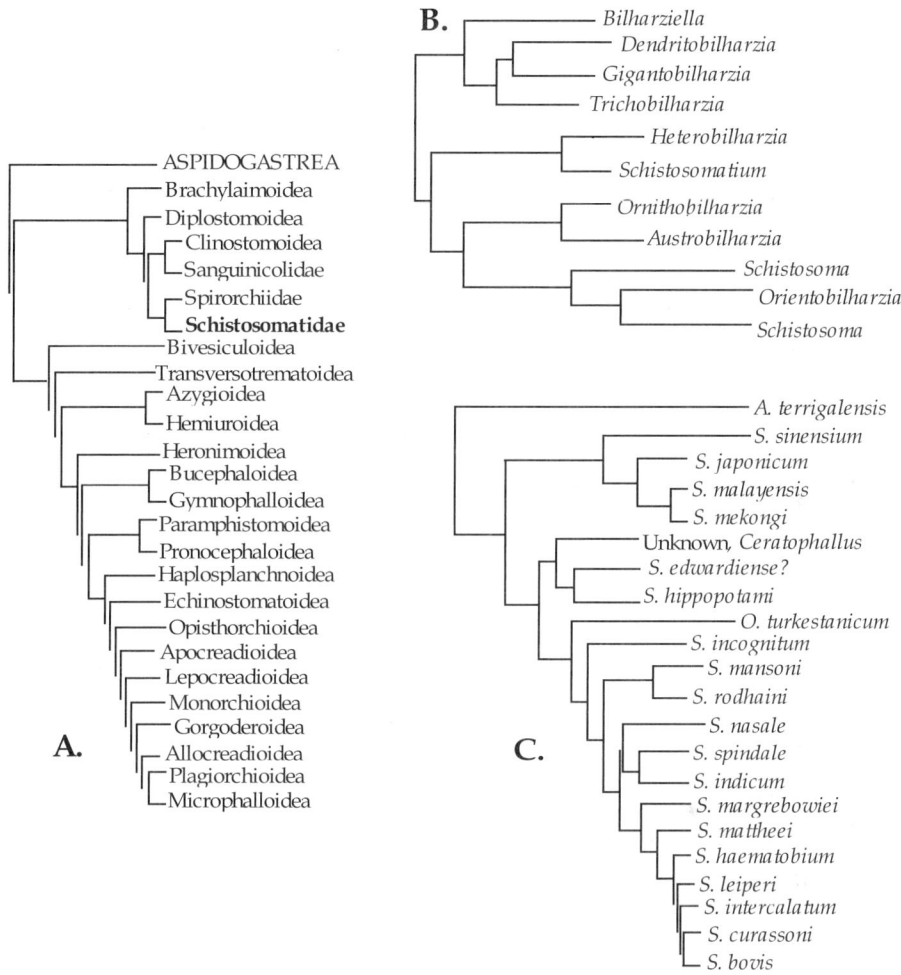

Figure 1-1. A. A summary of hypothesized relationships and higher classification of digenetic trematodes showing the relatively basal position of the schistosomes (Family Schistosomatidae), and that the turtle blood flukes (Family Spirorchiidae) comprise the likely sister group of the schistosomes. The phylogeny was estimated using the complete small subunit ribosomal RNA and partial large subunit ribosomal RNA gene sequences, based on data from Olson et al. (2003) and modified from Figure 3 of Cribb et al. (2003). B. A summary of relationships among the 10 genera of schistosomes for which sequence data are available, based on partial small and partial large subunit rRNA genes, ITS1, 5.8S, ITS2, mitochondrial cytochrome oxidase I and partial mitochondrial small subunit rRNA gene sequences, modified from Snyder and Loker (2000) and Lockyer et al. (2003). C. Relationships among the species of *Schistosoma*, using the same sequences identified in panel B, after Lockyer et al. (2003) and Morgan et al. (2003).

3. WHAT WAS THE PRIMORDIAL SCHISTOSOME LIKE?

The discovery by Platt et al. (1991) of the peculiar schistosome *Griphobilharzia amoena* from freshwater crocodiles in Australia provides a tantalizing glimpse into the schistosomes' past. This is the only species of schistosome known from other than a bird or mammal. It is also peculiar in that male and female worms seem to pair while still immature, with the female enclosed completely within a gynecophoric chamber of uncertain origin in the male (Platt et al., 1991). These authors raise the possibility that both sexes derive from a single cercaria. Unfortunately, we lack additional important information about this species. Although it is assumed that *G. amoena* must occupy a basal position within the Schistosomatidae, we lack any sequence data that could provide an independent corroboration of *G. amoena*'s phylogenetic position. This would also help us understand if the primordial schistosome was a parasite of crocodilians or other exothermic vertebrates, or if this family has exclusively colonized endotherms.

4. WHAT ARE THE FAMILY TIES OF *SCHISTOSOMA*?

We now have at least some sequence data for representatives of 10 of the 13 genera of schistosomes. We presently lack sequence information for *Griphobilharzia amoena*, the two known schistosome species from elephants (*Bivitellobilharzia*) and for *Macrobilharzia*, from cormorants and anhingas. So, another tangible objective for the future is to acquire sequence data for these genera so that a more complete and definitive phylogeny of the Schistosomatidae can be compiled. Regarding intrafamilial relationships, one reasonable hypothesis would be that the schistosomes are split into two major clades, one avian and one mammalian, but the results of two independent molecular studies (Snyder and Loker, 2000; Lockyer et al., 2003) suggest this is not the case (Fig. 1-1B). There is a basal clade comprised exclusively of four genera of avian schistosomes, but the remaining large clade can be subdivided into three smaller groups, with the North American mammalian schistosomes *Heterobilharzia* and *Schistosomatium* separated from the remaining mammalian schistosomes by the avian genera *Ornithobilharzia* and *Austrobilharzia*. The genus *Schistosoma*, which includes all the species that parasitize humans, is a derived lineage within the family.

Several studies have agreed that the one species thus far examined of *Orientobilharzia*, a genus of Asian mammalian schistosomes, nests within *Schistosoma* (Snyder and Loker, 2000; Zhang et al., 2001; Lockyer et al., 2003). Do the other two species of *Ornithobilharzia* also nest within *Schistosoma* in molecular phylogenies? Also, it would be particularly interesting to know how the elephant schistosomes fit into the picture. Do they represent a separate offshoot in the family or are they close relatives to one of the other lineages of schistosomes infecting mammals?

5.   WHAT ARE THE RELATIONSHIPS AMONG SPECIES WITHIN *SCHISTOSOMA*?

Traditionally, the 20 species of *Schistosoma* have been informally divided into four species groups (Table 1-1), with the groups based on geographic area of origin, egg shape and snail host. The *S. japonicum* group is exclusively Asian in its distribution, and its members are transmitted by pomatiopsid caenogastropod snails. Most representatives possess round eggs with rudimentary, recessed or absent spines. The *S. mansoni* group is known, or considered, to be transmitted by the planorbid snail genus *Biomphalaria*, and is primarily African in its distribution although *S. mansoni* occurs in both southwest Asia and South America. The *S. indicum* group is also Asian, is comprised of four species with eggs of variable shape and its members infect planorbid or lymnaeid pulmonate gastropods. The *S. haematobium* group with seven species is the largest of the four traditional species groups, is almost exclusively transmitted by the planorbid genus *Bulinus* and is mostly confined to Africa.

Table 1-1. The traditionally recognized species groups of *Schistosoma*.

S. japonicum group	*S. mansoni* group	*S. indicum* group	*S. haematobium* group
japonicum	mansoni	indicum	haematobium
mekongi	rodhaini	nasale	intercalatum
malayensis	hippopotami	spindale	bovis
sinensium	edwardiense	incognitum	mattheei
ovuncatum			curassoni
			margrebowiei
			leiperi

Do these four species groups accurately delineate relationships among species within *Schistosoma*, as assessed independently by molecular phylogenetics? A perusal of the hypothesized relationships among *Schistosoma* species as depicted in Figure 1-1C suggests that the answer is partially "yes" and partially "no". First, regarding the traditional *S. japonicum* group, several phylogenetic analyses (Snyder and Loker, 2000; Zhang et al., 2001; Agatsuma et al., 2001; Attwood et al., 2002; Lockyer et al., 2003) consistently affirm that the Asian pomatiopsid-transmitted species cluster together and that this lineage occupies a basal position in the genus. As noted by Agatsuma et al. (2001), and as recognized by other Asian schistosome workers (Attwood et al., 2002), *S. sinensium* and its close relative *S. ovuncatum* (and possibly other as yet undescribed species) are genetically distinct from the remaining species of the *S. japonicum* group. So, the Asian pomatiopsid-transmitted species (the traditional *S. japonicum* group) form a well-defined lineage that is comprised of two sublineages, *S. japonicum* and its allies and *S. sinensium* and its allies. Phylogenetic studies have also upheld the *Bulinus*-transmitted *S. haematobium* group, the species of which consistently cluster together and in this case occupy a more derived position in the tree.

The other two traditionally recognized species groups do not fare so well. For the *S. indicum* group, the three *Indoplanorbis*-transmitted species do in fact cluster together suggesting they comprise a natural group. Their close relationship to the *S. haematobium* group is of interest because their respective snail hosts, *Indoplanorbis* and *Bulinus,* are also close relatives. The fourth species of the *S. indicum* group, *S. incognitum,* occupies a much more basal position in the tree (Agatsuma et al., 2002; Lockyer et al., 2003). Interestingly, its placement on the tree is close to *Orientobilharzia turkestanicum,* a species with which it shares an Asian distribution and use of lymnaeid snail hosts.

The four species of the traditional *S. mansoni* group do not comprise a natural group and is essentially reduced to just two species (Fig. 1-1C), *S. mansoni* and *S. rodhaini.* The other two species, *S. hippopotami* and *S. edwardiense,* both parasites of the hippopotamus, and a third as yet undescribed species fall into a much more basal clade in the tree (Déspres et al., 1995; Morgan et al., 2003). Members of this newly recognized group are also distinguished by the presence of a long tail stem in the cercaria stage.

This same tree suggests humans or their immediate ancestors have been colonized by schistosomes on at least three and possibly as many as five separate occasions. Two members of the *S. japonicum* group infect humans, one of them being *S. japonicum* which is a relative generalist capable of infecting a broad range of mammals. The second species, *S. mekongi,* commonly infects dogs and humans. *S. mansoni* is today predominantly a human schistosome though it can and does infect rodents and its origins are likely as a rodent parasite that subsequently adapted to humans (Combes, 1990). *S. haematobium* and its close relative *S. intercalatum* probably originated in ungulates and separately colonized humans (Combes, 1990).

The former species with its distinctive site of egg deposition in the urinary system is the most human-adapted of all the *Schistosoma* species. We can conclude from the broad pattern of human use on the *Schistosoma* family tree that schistosomes readily colonize humans and that human-borne schistosomes are today faring very well. By virtue of being long-lived, large-bodied, mobile, water-loving and slow to develop immunity, humans provide a stable and productive environment in which the worms can propagate and be disseminated to their snail hosts.

6.   WHERE DID *SCHISTOSOMA* COME FROM?

Africa currently supports a large number of successful *Schistosoma* species so it is not unreasonable to postulate that the genus originated there, and subsequently spread to other continents (Davis, 1980, 1992). However, molecular phylogenetic studies consistently retrieve a basal position of the exclusively Asian *S. japonicum* group in the *Schistosoma* tree, suggestive of an Asian origin (Snyder and Loker, 2000; Zhang et al., 2001; Morgan et al., 2001; Agatsuma et al., 2001, 2002; Attwood et al., 2002; Lockyer et al., 2003). Remarkably, the order of the genes in the mitochondrial genome for members of the lineage containing *S. japonicum* are like those of other flatworms, whereas the mitochondrial gene order for primarily African species such as *S. mansoni* and *S. haematobium* are different, further suggesting the African *Schistosoma* species lie in a more derived position (Le et al., 2001; Lockyer et al., 2003). Studies of C-banding chromosomal patterns by Hirai et al. (2000) also are concordant with the view that *S. japonicum* represents a basal position within the genus. Thus although many of the extant species of *Schistosoma* are African and the genus assumes its greatest medical and veterinary significance there, three lines of evidence suggest its origins lie in Asia.

The relatively basal lineage recently identified by Morgan et al. (2003) is comprised of species only known to be in Africa and their snail hosts are all genera that are found exclusively, or largely, within sub-Saharan Africa. Although this would strongly imply the origin of this clade is African, the fact that hippos once radiated extensively into Eurasia might suggest that the parasites were acquired there and brought back to Africa. In general as one ascends the topology of the tree, species that are Asian tend to alternate with groups that are African suggesting there has been a complicated series of movements of *Schistosoma* between continents. One example of the complexities involved is provided by the *S. indicum* group. Did members of this group arise in Africa and later colonize Asia (Barker and Blair, 1996), or were they originally in Asia and give rise to African species of the *S.*

haematobium group? Given that the clade basal to the *S. indicum* group is comprised only of African members, we would argue for the former scenario.

7.   WHAT DO THE SNAILS HAVE TO TELL US?

Insofar as schistosomes are totally dependent on snails for completion of their life cycles, and their present day distributions are in general more limited by their snail hosts than their vertebrate hosts, it is important to gain some understanding of the histories and relationships among the snails that serve as intermediate hosts. Within *Schistosoma*, the five species of the basal *S. japonicum* group are all transmitted by operculate snails of the family Pomatiopsidae whereas the remaining species are transmitted by basommatophoran pulmonate snails of two families, the Lymnaeidae and the Planorbidae. It is clear that pomatiopsids and the two pulmonate families are not close relatives (Blair et al., 2001) so unless we are missing a great many representatives of extinct *Schistosoma* that lived in some of the intermediate snail groups, one of the defining events early in the evolution of *Schistosoma* must have been a dramatic snail host shift. If pomatiopsid-transmitted *Schistosoma* are indeed basal, then the shift would have been from pomatiopsid to basommatophoran snail. Once in basommatophorans, to account for the known host usage patterns, the parasites must have jumped to new snail hosts on several different occasions. The lineage containing the hippo parasites perhaps best exemplifies this trend as each of the three species develops in a different genus of snail host: *S. hippopotami* in *Bulinus*, putative *S. edwardiense* in *Biomphalaria*, and the third undescribed species in *Ceratophallus*, a snail genus not previously known to support *Schistosoma* sporocyst development. Furthermore, these snails are not particularly close relatives: *Bulinus* is basal and *Biomphalaria* derived within the Planorbidae (Morgan et al., 2002). So we are left with a bit of a paradox: although present-day schistosomes are unquestionably specific with respect to the snail hosts they currently use (for example, *S. mansoni* can only develop in *Biomphalaria* and *S. haematobium* can only develop in *Bulinus*), their history is replete with examples of host shifts. We don't understand how such shifts occurred.

The potential importance of understanding the evolutionary history of the snail hosts is exemplified by *Biomphalaria*. Whereas *Biomphalaria* was once considered to have been present in Africa before its separation from South America some 70 million years ago, recent phylogenetic studies (see references in DeJong et al., 2001), backed up by studies of the fossil record, suggest that *Biomphalaria* originated in South America and secondarily colonized Africa, within the past five million years. This in turn implies that *S. mansoni* and its sister *species S. rodhaini*, both of which are *Biomphalaria*-transmitted, could not have existed as we know them today before *Biomphalaria's* introduction to Africa. Based purely on this reasoning, the origin of the *Biomphalaria*-transmitted human parasite *S. mansoni* must have been within this time frame as well. This provides a new point of view to anyone attempting to reconstruct how schistosomes may have interacted with evolving hominids in Africa. There is still much to do with respect to unraveling snail phylogenies. As just one example, *Biomphalaria salinarum*, a species indigenous to southwest Africa, has yet to be investigated with modern methods. Such a study would help us to understand the history of *Biomphalaria* in Africa.

8. SOME FINAL THOUGHTS – WHAT'S PAST IS PROLOGUE

Ongoing efforts to reconstruct the evolutionary history of schistosomes and their snail hosts have provided some notable new insights about basic schistosome biology – their relatively basal status among all digeneans, a better understanding of relationships among *Schistosoma* species, the facility with which schistosomes have shifted snail hosts over the millennia, new views about where *Schistosoma* originated and the possible ages of some of the prominent schistosomes of humans. As noted above, much exciting work remains to be done. Having solid hypotheses regarding their evolutionary past will help us understand some of the momentous changes lying ahead with respect to schistosomes. The wide scope of the Gates Foundation-funded schistosomiasis control initiative will no doubt impact the evolution of schistosomes possibly in ways quite unforeseen. Human-mediated environmental impacts will continue to greatly affect the distribution of snails that serve as hosts. The HIV pandemic, by creating immunosuppressed and extraordinarily susceptible hosts, might also influence the course of schistosome evolution (Combes and Jourdane, 1991). In each of these cases, knowing where the schistosomes have been should help us figure out where they are going.

ACKNOWLEDGMENTS

The authors thank Dr. Jess A. T. Morgan and Dr. Randy J. DeJong for their assistance in preparation of this manuscript, and Dr. T.H. Cribb and Dr.

D.T.J. Littlewood for their help in providing figures. We thank several collaborators from around the world who have provided us specimens of both worms and snails that helped to build the conclusions mentioned above. This work was supported by NIH grant RO1 AI44913.

REFERENCES

Agatsuma, T., Iwagami, M., Liu, C.X., Rajapakse, R.P.V.J., Mondal, M.M.H., Kitikoon, V., Ambu, S. Agatsuma, Y., Blair, D., Higuchi, T., 2002, Affinities between Asian non-human *Schistosoma* species, the *S. indicum* group, and the African human schistosomes, *J. Helminthol.* **76:**7-19.

Agatsuma, T., Iwagami, M., Liu, C.X., Saitoh, Y., Kawanaka, M., Upatham, S., Qui, D. and Higuchi, T., 2001, Molecular phylogenetic position of *Schistosoma sinensium* in the genus *Schistosoma*, *J. Helminthol.* **75:**215-221.

Attwood, S.W., Upatham, E.S., Meng, X.H., Qiu, D.C., Southgate, V.R., 2002, The phylogeography of Asian *Schistosoma* (Trematoda: Schistosomatidae), *Parasitology* **125:** 99-112.

Barker, S.C., and Blair, D., 1996, Molecular phylogeny of *Schistosoma* species support traditional groupings within the genus, *J. Parasitol.* **82:**292-298.

Blair, D., Davis, G.M., and Wu, B., 2001, Evolutionary relationships between trematodes and snails emphasizing schistosomes and paragonimids, *Parasitology* **123:**S229-S243.

Brooks, D.R., O'Grady, R.T., and Glen, D.R., 1985, Phylogenetic analysis of the Digenea (Platyhelminthes, Cercomeria) with comments on their adaptive radiation, *Can. J. Zool.* **63:**411-443.

Carmichael, A.C., 1984, Phylogeny and historical biogeography of the Schistosomatidae. Ph.D. Dissertation, Michigan State University. pp 224.

Chitsulo, L., Engels, D., Montresor, A., and Savioli, L., 2000, The global status of schistosomiasis and its control, *Acta Tropica* **77:** 41-51.

Combes, C., 1990, Where do human schistosomes come from? An evolutionary approach. *Trends Ecol. Evol.* **5:**334-337.

Combes, C., and Jourdane, J., 1991, Immunodeficieincies in humans – a possible cause of hybridizations and gene introgressions in bisexual parasites, *Acta Oecologica – Int. J. Ecol.* **12:**829-830.

Cribb, T.H., Bray, R.A., Olson, P.D., and Littlewood, D.T.J., 2003, Life cycle evolution in the Digenea: a new perspective from phylogeny, *Adv. Parasitol.* (in press).

Davis, G.M., 1980, Snail hosts of Asian *Schistosoma* infecting man: Evolution and coevolution, in: *The Mekong Schistosome*, J. Bruce, and S. Sornmani, eds., Malacological Review, Michigan, USA, pp. 195-238.

Davis, G.M., 1992, Evolution of Prosobranch snails transmitting Asian *Schistosoma*, coevolution with *Schistosoma*: A review, *Prog. Clin. Parasitol.* **3:**145-204.

DeBont, J., and Vercruysse, J., 1998, Schistosomiasis in cattle, *Adv. Parasitol.* **41:**285-364.

DeJong, R.J., Morgan, J.A.T., Paraense, W.L., Pointier, J-P., Amarista, M., Ayeh-Kumi. P.F.K., Babiker, A., Barbosa, C.S., Bremond, P., Canese, A.P., de Souza, C.P., Dominguez, C., File, S., Gutierrez, A., Incani, R.N., Kawano, T., Kazibwe, F., Kpikpi, J., Lwambo, N.J.S., Mimpfoundi, R., Poda, J-N., Sene, M., Velasquez, L.E., Yong, M., Adema, C.M., Hofkin, B.V., Mkoji, G.M. and Loker, E.S., 2001, Evolutionary relationships and biogeography of

Biomphalaria (Gastropoda: Planorbidae) with implications regarding its role as host of the human bloodfluke, *Schistosoma mansoni. Mol. Biol. Evol.* **18**:2225-2239.

Després, L., Kruger, F. J., Imbert-Establet, D., and Adamson, M. L., 1995, ITS2 ribosomal RNA indicates *Schistosoma hippopotami* is a distinct species, *Int. J. Parasitol.* **25**:1509-1514.

Hirai, H., Taguchi, T., Saitoh, Y., Kawanaka, M., Sugiyama, H, Habe, S., Okamoto, M., Hirata, M., Shimada, M., Tiu, W.U., Lai, K., Upatham, E.S. and Agatsuma, T., 2000, Chromosomal differentiation of the *Schistosoma japonicum* complex, *Int. J. Parasitol.* **30**: 441-452.

Le, T.H., Humair, P.F., Blair, D., Agatsuma, T., Littlewood, D.T.J. and McManus, D.P., 2001, Mitochondrial gene content, arrangement and composition compared in African and Asian schistosomes, *Mol. Biochem. Parasitol.* **117**: 61-71.

Lockyer, A. E., Olson, P. D., Østergaard, P., Rollinson, D., Johnston, D. A., Attwood, S. W., Southgate V. R., Horak, P., Snyder, S. D., Le, T. H., Agatsuma, T., McManus, D. P., Carmichael, A. C., Naem, S., and Littlewood, D. T. J., 2003, The phylogeny of the Schistosomatidae based on three genes with emphasis on the interrelationships of *Schistosoma* Weinland, 1858, *Parasitology* **126**:203-224.

McCully, R., Niekerk, J., and Kruger, S., 1967, Observations on the pathology of bilharziasis and other parasitic infestations of *Hippopotamus amphibius* Linnaeus, 1758, from the Kruger National Park, *Onderstepoort J. Vet. Res.* **34**: 563-618.

McLaren, D.J., and Hockley, D.J., 1977, Blood flukes have a double outer membrane, *Nature* **269**:147-149.

Morgan, J.A.T., DeJong, R.J., Jung, Y., Khallaayoune, K., Kock, S., Mkoji, G.M., and Loker, E.S., 2002, A phylogeny of planorbid snails, with implications for the evolution of *Schistosoma* parasites, *Mol. Phylogenet. Evol.* **25**:477-488.

Morgan, J.A.T., DeJong, R.J., Kazibwe, F., Mkoji, G.M. and Loker, E.S., 2003, A new lineage of *Schistosoma. Int. J. Parasitol.* **33**:977-985.

Morgan, J.A.T., DeJong, R.J., Snyder, S.D., Mkoji, G.M. and Loker, E.S. 2001. *Schistosoma mansoni* and *Biomphalaria*: past history and future trends. Parasitology 123: S211-S228.

Olson, P.D., Cribb, T.H., Tkach, V.V., Bray, R.A., and Littlewood, D.T.J., 2003, Phylogeny and classification of the Digenea (Platyhelminthes:Trematoda), *Int. J. Parasitol.* (in press).

Platt, T.R., Blair, D., Purdie, J., and Melville, L., 1991, *Griphobilharzia amoena* n. gen, n. sp. (Digenea, Schistosomatidae), a parasite of the freshwater crocodile *Crocodylus johnstoni* (Reptilia, Crocodylia) from Australia, with the erection of a new subfamily, Griphobilharziinae, *J. Parasitol.* **77**:65-68.

Platt, T.R., and Brooks, D.R., 1997, Evolution of the schistosomes (Digenea: Schistosomatoidea): The origin of dioecy and colonization of the venous system. *J. Parasitol.* **83**:1035-1044.

Snyder, S. D., and Loker, E. S., 2000, Evolutionary relationships among the Schistosomatidae (Platyhelminthes: Digenea) and an Asian origin for *Schistosoma, J. Parasitol.* **86**:283-288.

Ulmer, M.J., and Vande Vusse, F.J., 1970, Morphology of *Dendritobilharzia pulverulenta* (Braun, 1901) with notes on secondary hermaphroditism in males, *J. Parasitol.* **56**:67-74.

Zhang, G.J., Verneau, O., Qiu, C.P., Jourdane, J., and Xia, M.Y., 2001, An African or Asian evolutionary origin for human schistosomes? *C. R. Acad. Sci. III* **324**:1001-1010.

Chapter 2

SCHISTOSOME RETROTRANSPOSONS

Paul J. Brindley,[1,2] Claudia S. Copeland,[1,2] and Bernd H. Kalinna[3]

[1]Department of Tropical Medicine and [2]Interdisciplinary Program in Molecular and Cellular Biology, Tulane University Health Sciences Center, New Orleans, Louisiana 70112; [3]Department of Molecular Parasitology, Institute for Biology, Humboldt University Berlin, 10155 Berlin, Germany

Key words: Schistosome; genome; mobile genetic elements; retrotransposon; reverse transcriptase

1.   INTRODUCTION

Eukaryotic genomes generally contain substantial amounts of repetitive sequences, many of which are mobile genetic elements (e.g., Lander *et al.*, 2001; Holt *et al.*, 2002). These mobile sequences have played fundamental roles in the evolution of the human and other eukaryotic genomes (Charlesworth *et al.*, 1994; Deininger and Batzer, 2002), and are among the most powerful endogenous human mutagens (Kazazian, 1999; Dewannieux *et al.*, 2003). Although less is known about the schistosome genome, recent findings suggest that up to half of the entire schistosome genome may be comprised of repetitive sequences, and much of this repetitive complement will be comprised of mobile genetic elements (see Brindley *et al.*, 2003). Here we review a series of mobile genetic elements from the schistosome genome, focusing on schistosome retrotransposable sequences. The identity, structure, phylogenetic relationships, and contribution of these elements to genome size in schistosomes are described, and we address their probable role in schistosome evolution and potential utility in introducing transgenes into schistosomes and other applications.

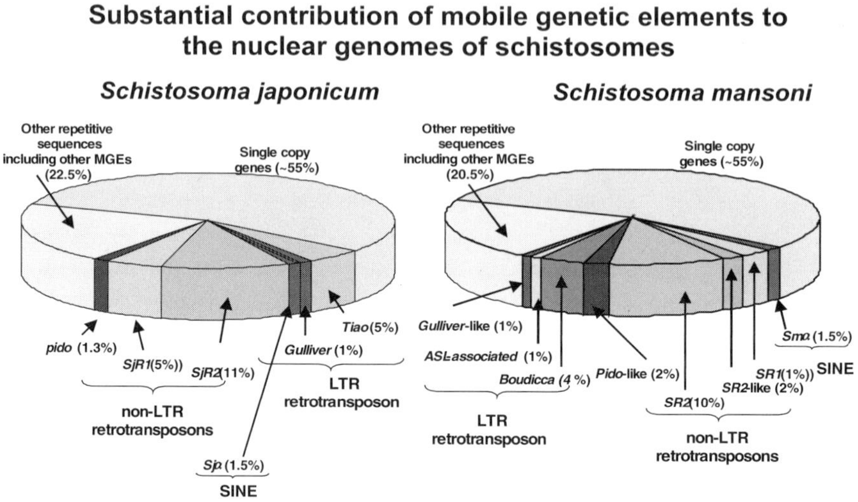

Figure 2-1. Predicted contribution of mobile genetic elements to the nuclear genomes of *Schistosoma japonicum* and *S. mansoni* are represented in pie charts, predicted from published and gene database sources (Adapted from Brindley et al., 2003, with permission).

2.   THE SCHISTOSOME GENOME

Schistosomes have a comparatively large genome, estimated at ~270 megabase pairs for the haploid genome of *Schistosoma mansoni*, arrayed on seven pairs of autosomes and one pair of sex chromosomes (Simpson *et al.*, 1982). For comparison, the schistosome genome is about the same size as that of the puffer fish, *Fugu rubripes*, two to three times the size of that of the angiosperm, *Arabidopsis thaliana*, or the free-living nematode, *Caenorhabditis elegans*, ten times the size of the *Plasmodium falciparum* genome, and about one tenth the size of the human genome. The other major schistosome species parasitizing humans probably have a genome of similar size to that of *S. mansoni*, based on their karyotypes (Hirai *et al.*, 2000). Though none of the schistosome genomes have been sequenced in their entirety, several hundred thousand schistosome expressed sequence tags (ESTs) and genome survey sequences have been lodged in GenBank, probably covering the entire transcriptome and indicating that there are ~14,000 genes in *S. mansoni* (Verjovski-Almeida *et al.*, 2003; Hu *et al.*, 2003). The mobile genetic elements (MGEs) of the schistosome genome include SINE-like elements, non-long terminal repeat (non-LTR) retrotransposons and LTR retrotransposons (Figs. 2-1-2-4), and appear to make up at least one quarter of the schistosome genome.

3. CATEGORIES OF TRANSPOSABLE ELEMENTS AND MODES OF TRANSPOSITION

MGEs are grouped in two major categories, Class I and Class II (Finnegan, 1992). Class I elements transpose through a RNA intermediate whereas Class II elements transpose directly as DNA. Class I comprises (a) the long terminal repeat (LTR) retrotransposons and the retroviruses, (b) the non-LTR retrotransposons, and (c) the short interspersed nuclear elements (SINEs). Class I elements occur in taxa as diverse as fungi and mammals, and are mobilized by replicative processes that generate numerous daughter copies and facilitate insertion into the host genome, thereby directly expanding the size of the host genome. Class II elements are termed 'transposons', and include groups from prokaryotes and eukaryotes.

LTR retrotransposons resemble retroviruses in their structure and intracellular life cycles. These elements are typically 5 - 10 kb in length. Their general structure consists of two open reading frames (ORFs) flanked by long direct terminal repeats of ~200-600 bp in length (Fig. 2-2). Some, such as *gypsy* from *Drosophila melanogaster*, *Osvaldo* from *Drosophila buzzatii*, and *Tas* from *Ascaris lumbricoides*, include a third ORF, *env*, encoding the envelope protein characteristic of retroviruses (Fig. 2-2). Retroviruses probably evolved from LTR retrotransposons, mediated by the acquisition of envelope proteins that facilitated extracellular existence and horizontal transmission between cells and species (Malik *et al.*, 2000). The LTRs play a pivotal role in initiating transcription and in transposition. The first ORF, *gag*, encodes a polyprotein precursor that is later processed to yield the structural proteins making up the virion core. Of these, the nucleocapsid protein associates directly with the RNA, and exhibits a characteristic cysteine/histidine motif, which appears to function as a zinc finger domain. The second ORF, *pol*, encodes a polyprotein with discrete protease, reverse transcriptase (RT), RNaseH, and integrase enzyme domains. The *pol* domain order varies between the two major *gypsy/Ty3* and *Copia/Ty1* clades of LTR retrotransposons. In retroviruses and LTR retrotransposons with an *env* gene, the envelope protein associates with the cell membrane, which envelops the virion core, allowing the viral particle to bud off from the host cell. Envelope facilitates infection via attachment to specific cell surface receptors. Thus, in addition to vertical transmission in the germ line, LTR retrotransposons with *envelope* genes are capable of extracellular existence and horizontal transmission.

Mobile genetic elements that transpose via RNA intermediates

LTR Retrotransposons and Retroviruses

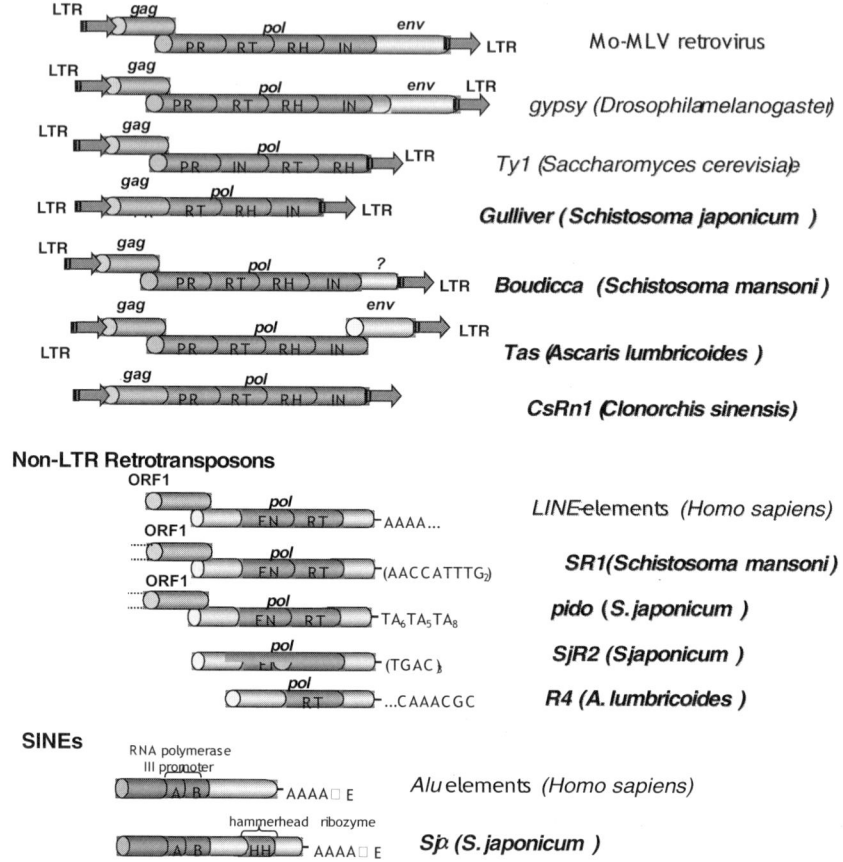

Figure 2-2. Schematic representation of the structure of representative retrotransposable elements from schistosomes and other parasitic helminths (denoted in bold text) and other hosts. Abbreviations: gag, group associated antigen; pol, polyprotein; env, envelope; PR, protease, RT, reverse transcriptase; RH, RNaseH, IN, integrase; EN, endonuclease; LTR, long terminal repeat (Adapted from Brindley et al., 2003, with permission).

Non-LTR retrotransposons are usually ~ 4 - 6 kb in length, generally have two ORFs, often have A-rich 3'-termini, and are transmitted vertically. The Long Interspersed Nuclear Elements (LINEs) of humans are well known members. Full length LINE1 is bicistronic: the product of the first ORF has RNA binding function, although this is not well characterized, whereas the second ORF encodes a polyprotein with RT and apurinic endonuclease (APE) activities. Of 11 clades of non-LTR retrotransposons recognized, several also have an RNaseH domain within the polyprotein (Malik *et al.*, 1999). The movement of these elements involves binding of the ORF2 product to both the full-length retrotransposon RNA transcript and the host chromosome at the integration target site. Nicking of the host chromosome by APE activity of the ORF2 product provides an initiation start site, a 3'-hydroxyl group, to prime reverse transcription of the retrotransposon RNA. Target site duplication, a footprint of retrotransposition, is a by-product of integration of non-LTR retrotransposons into the host chromosome. Most copies of non-LTR retrotransposons are 5'-truncated, the consequence of premature termination of reverse transcription (see Deininger and Batzer, 2002).

SINEs are short (< 600 bp in length), non-autonomous MGEs, with poly A-rich 3'-termini (like most non-LTR retrotransposons) (Fig. 2-2). Most SINEs have a composite structure comprising a 5' tRNA-related region followed by a tRNA-unrelated region. *Alu*'s, however, which are the best characterized SINEs and which are known only from humans and some other primates, have evolved from the 7SL RNA gene (Deininger and Batzer, 2002). SINEs do not encode any proteins of their own, but rely on RT from other sources for their mobilization. Non-LTR retrotransposons provide the RT activity that drives retrotransposition of SINEs (Kajikawa and Okada, 2002; Dewannieux *et al.*, 2003). Transcription of SINEs is driven by RNA polymerase III followed by reverse transcription and integration into the new site of the genome. SINE transcripts appear to be incorporated into the ribonucleoprotein particles that are the transposition intermediates for non-LTR retrotransposons, from where they are transported into the nucleus, reverse transcribed, and integrated into the host chromosome. SINEs are transmitted vertically.

Transposons constitute the Class II MGEs. They move by a 'cut and paste' process that is generally independent of host-specific factors. Transposons range in size from ~ 1.3 to 3.0 kb and are bounded by terminal inverted repeat (IR) sequences that are recognized by the transposase enzyme, the only protein encoded by the transposon. Transposase mediates excision of the transposon and its re-insertion into the host genome. Transposons can move horizontally between species, as well as being propagated through the germ line. In evolutionary terms, transposons require horizontal movement between species for their long-term survival. (This attribute has been pivotal in harnessing of transposons such as *mariner* as vectors for transgenesis.) Because transposase is *trans*-acting, cutting and pasting both older and younger copies of the transposon, and because older copies tend to have accumulated deleterious mutations, efficiency of transposition declines over time until extinction.

4.   SCHISTOSOME RETROTRANSPOSABLE ELEMENTS

4.1   *Boudicca*

Boudicca is a ~6 kb long terminal repeat retrotransposon from *S. mansoni*. Two 328bp LTRs flank a coding region consisting of open reading frames representing the *gag* and *pol* polyproteins, 5' and 3' untranslated regions, and, at least in the well characterized copy, an additional unknown third open reading frame (Copeland *et al.*, 2003). *Boudicca* is a high copy number element, estimated at 2,000 to 3,000 copies per haploid genome, and is actively transcribed in adult worms, cercariae, and sporocysts. One of the only members of the *Kabuki/CsRn1* clade of gypsy-type LTR retrotransposons, *Boudicca*'s closest relatives are *PwRn1* from *Paragonimus westermani*, *CsRn1* from *Clonorchis sinensis*, and *Kabuki* from the silkworm, *Bombyx mori*. This clade, closely related to the errantiviruses, is differentiated not only on the primary sequence level but also by a unique Cys-His box structure, CHCC instead of the more common retroviral CCHC Cys-His box (Bae *et al.*, 2002). Currently, we are investigating the potential of *Boudicca* as a transgenesis vector for schistosomes.

4.2   *Gulliver*

The LTR retrotransposon *Gulliver*, ~4.8 kb in length, is present in multiple copies in the genome of *S. japonicum* (Figs. 2-1, 2-2) (Laha *et al.*, 2001). Southern blot analysis indicates that it is also present in the genome of *S. mansoni*. The LTRs of *Gulliver* are 259 bp in length and include RNA polymerase II promoter sequences, a CAAT signal and a TATA box. *Gulliver* exhibits features characteristic of a functional LTR retrotransposon including two read through (termination) ORFs encoding retroviral gag and pol proteins of 312 and 1,071 amino acid residues, respectively. The *gag* ORF encodes motifs conserved in nucleic acid binding proteins, while the *pol* ORF encodes conserved domains of aspartic protease, reverse transcriptase (RT), RNaseH and integrase, in that order, a pol pattern conserved in the *gypsy* lineage of LTR retrotransposons. Its structure is similar to that of *gypsy*, although *Gulliver*'s closest relatives include *mag* from *B. mori* and *Blastopia* from *D. melanogaster* (Figs. 2-2, 2-3).

Phylogenetic relationships of LTR retrotransposons of schistosomes

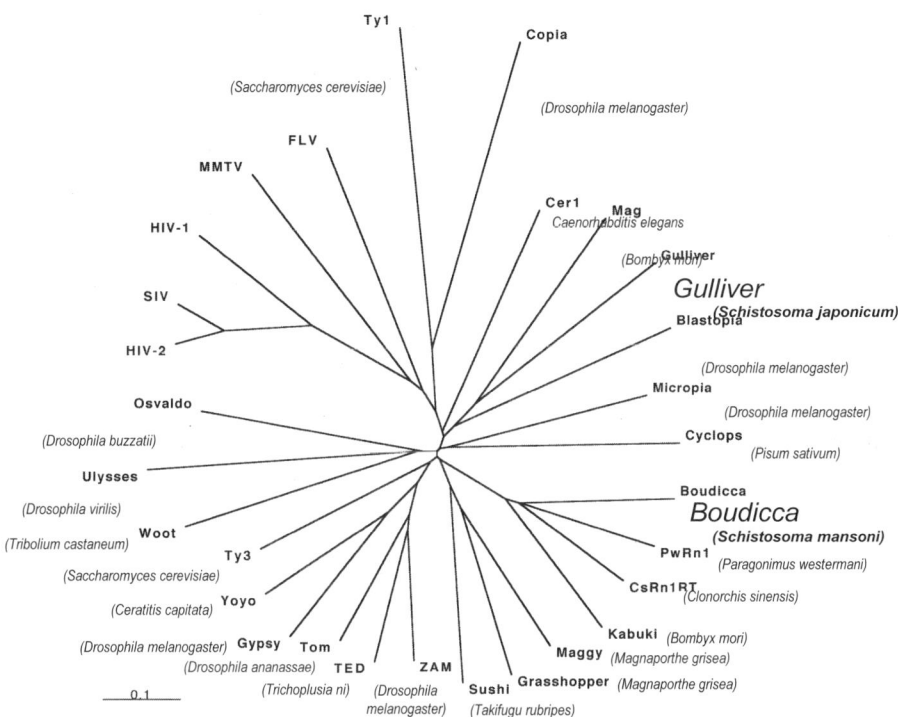

Figure 2-3. Phylogenetic tree comparing the relationships among the schistosome LTR retrotransposons Boudicca, Gulliver and other LTR retrotransposons and retroviruses. Reverse transcriptase regions were aligned and tree files were created using MacVector and ClustalX software, and unrooted output trees were created using Treeview software. Accession numbers: Blastopia: CAA81643, Boudicca: AY308018-AY308026, Cer1: AAA50456, Copia: OFFFCP, CsRn1: AAK07486, Cyclops: AAL06415, HIV-1: P04585, HIV-2: AAA76841, FLV: NP_047255, Grasshopper: AAA21442, Gulliver: AF243513, Gypsy: GNFFG1, Kabuki: BAA92689, Mag: S08405, Maggy: AAA33420, Micropia: S02021, MMTV: GNMVMM, Osvaldo: CAB39733, PwRn1: AY237162, SIV: AAA47606, Sushi: AAC33526, TED: AAA92249, Tom: CAA80824, Ty1: P47100, Ty3: AAA35184, Ulysses: CAA39967, Woot: AAC47271, Yoyo: T43046, Zam: CAA04050.

4.3   Other LTR retrotransposons

Tiao, the first LTR type retrotransposon to be reported from any of the human schistosomes, is a high copy number (about 10,000) LTR retrotransposon found in *S. japonicum* (Genbank AF073334) (Fan and Brindley, 1998).   *Tiao* is related to the *Pao*-like retrotransposons *Kamikaze,*

Yamato, and *Pao* from *B. mori*, and *Ninja, BEL*, and *Max* from *Drosophila melanogaster*.   In *S. mansoni*, fragments of additional LTR type retrotransposons distinct from *Boudicca* have also been identified. Within the locus of the *S. mansoni* gene encoding adenylosuccinate lyase, intron 6 contains two 500 bp direct LTRs flanking a 1.9 kb sequence with homology to LTR retrotransposon type gag and pol (Foulk *et al.*, 2002). In addition, several other LTR type retrotransposons related to *Pao, Mag*, and *Osvaldo* are present in the genome of *S. mansoni*, awaiting characterization (Copeland and Brindley, unpublished).

4.4   The non-LTR retrotransposon *SR1*

SR1 is a non-LTR retrotransposon and was the first autonomous retrotransposon to be reported from schistosomes (Drew and Brindley, 1997). Orthologous versions occur in *S. mansoni (SR1)* and *S. japonicum (SjR1)*, with a copy number estimated in the range of 200 to 2,000 (Fig. 2-1). *SR1* is a *CR1*-like element (Figs. 2-2, 2-4), although the full-length element (predicted to be ~ 4 kb in length) has not been totally characterized. *SR1* elements possess atypical 3' termini consisting of the tandem repeat (AACCATTTG)$_2$ which are similar in structure to the imperfect tandem repeat of the 3' termini of *CR1* from chickens and other vertebrates.

4.5   *SR2*

SR2 is a non-LTR retrotransposon of ~3.9 kb in length with a single ORF encoding APE and RT. Orthologous versions appear to occur in *S. mansoni* (*SR2*) and *S. japonicum* (*SjR2*), with copy numbers estimated in the range of 1,000 to 10,000 (Drew *et al.*, 1999; Laha *et al.*, 2002a). *SR2* elements are related to *RTE-1* from *C. elegans*. The ORF is bounded by 5'- and 3'-terminal untranslated regions and, at its 3'-terminus, *SjR2* bears a short (TGAC)$_3$ repeat (Figs. 2-1, 2-2, 2-4).   Active, recombinant RT of *SjR2* has been produced in insect cells (Laha *et al.*, 2002a).

4.6   *pido*

This is the third non-LTR retrotransposon characterized from the genome of *S. japonicum* (Laha *et al.*, 2002b). Although the full-length element has yet to be characterized, a consensus sequence of 3564 bp of the truncated *pido* element has been assembled from several genomic fragments that contained *pido*-hybridizing sequences. The sequence encodes part of the first ORF, the entire second ORF and, at its 3'-terminus, a tandemly repetitive,

A-rich ($TA_6TA_5TA_8$) tail (Fig. 2-2). ORF1 of *pido* encodes a nucleic acid binding protein and ORF2 encodes a retroviral-like polyprotein with APE and RT domains. mRNA encoding the RT of *pido* was detected by reverse transcription-PCR in the egg, miracidium and adult developmental stages of *S. japonicum*, indicating that the RT domain is transcribed and suggesting that *pido* is replicating actively and mobile within the *S. japonicum* genome. At least 1,000 partial copies of *pido*, which is not closely related to either SR1 or SR2 (Figs. 2-1, 2-4), are dispersed throughout the genome of *S. japonicum*.

Phylogenetic relationships of Non-LTR retrotransposons of schistosomes

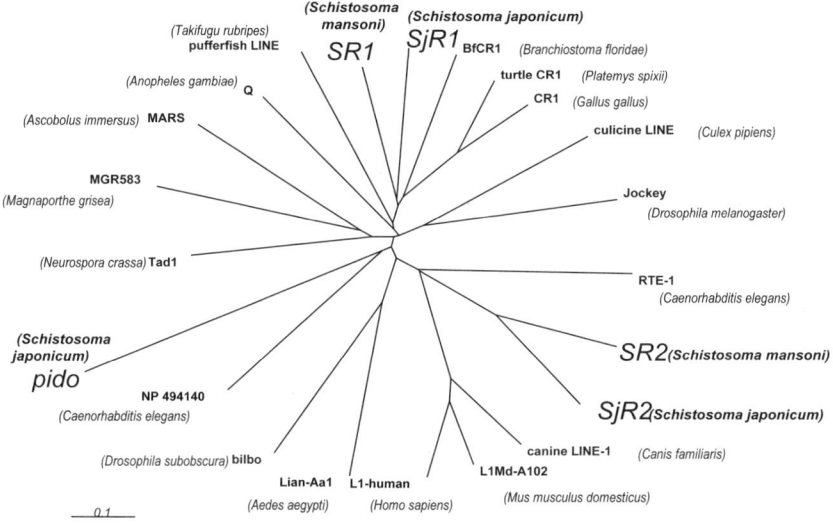

Figure 2-4. Phylogenetic tree comparing the relationships among the schistosome non-LTR retrotransposons SR1, SR2 and pido and other non-LTR retrotransposons. Reverse transcriptase regions were aligned and tree files were created using MacVector and ClustalX software, and unrooted output trees were created using Treeview software. Accession numbers: BfCR1: AAL40415, Bilbo: AAB92389, canine LINE-1: BAA25253, Cr1: AAC60281, culicine LINE: AAA28291, Jockey: P21328, L1-human: AAD04635, L1Md-A102: AAL17970, Lian-Aa1: T30319, MARS: CAA67543, MGR583: AAB71689, NP 494140: NP_494140, Pido: AY034006, pufferfish LINE: AAD19348, Q: T43020, RTE-1: S68633, SjR1: AAC62955, SjR2: AAK14815, SR1: AAC06264, SR2: AAC24982, Tad1: AAA21781, turtle CR1: BAA88337.

4.7   Sm•-, Sj•-like SINEs

SINE-like retroposons known as Sm , Sj , Sh etc occur in high copy number, 7000-10,000 copies, in the genomes of schistosomes (Spotila *et al.*, 1989; Drew and Brindley, 1995; Ferbeyre *et al.*, 1998; Laha *et al.*, 2000). The sequence of the consensus Sm –like SINE includes the hallmark features of SINE-like elements including a promoter region for RNA polymerase III, a region of identity of the bovine tRNA^{arg} gene, an AT-rich stretch at its 3'-terminus, a short length of 330 bp or less, and short direct repeat sequences flanking the insertion site (Figs. 2-1, 2-2). Sm encodes an active ribozyme bearing a hammerhead domain; Ferbeyre *et al.* (1998) suggested a role for these self-catalytic RNAs in control of retrotransposon expansion. Some Sm elements appear to be linked to the female (W) sex chromosome (Drew and Brindley, 1995). Based on the interaction of *Alu* elements and other SINES with LINEs (Dewannieux *et al.*, 2003), the RT activities of *pido*, *SR2* and/or *SR1* are likely utilized by the Sm –like elements for replication and dispersal.

5.   (RETRO)TRANSPOSONS FOR SCHISTOSOME TRANSGENESIS?

The ability to introduce transgenes into genomes of pathogenic organisms has revolutionized molecular approaches to microbial diseases. Whereas this revolution has included protozoan parasites, and indeed mosquito and other insect vectors, so far it has largely bypassed medical helminthology because neither cell lines nor transgenesis systems have been developed for parasitic worms. Consequently, molecular research in schistosomes and schistosomiasis, and indeed other helminthoses, has been structurally disadvantaged in comparison with other microbes (Boyle and Yoshino, 2003).   Since MGEs are valued as vectors for genetic transformation of other invertebrate genomes (e.g., O'Brochta *et al.*, 2003), likewise they hold promise for genetic transformation of schistosomes. Many free-living taxa are malleable to transformation with transposons. Because transposons often exhibit minimal requirements in terms of host cell factors for mobility, they can mobilize in genomes of diverse species (Guerios-Filho and Beverley, 1997; O'Brochta *et al.*, 2003). There are no reports of transformation of schistosomes or indeed other parasitic helminths with transposons, although this should be feasible now that conditions for transgenesis are being elucidated (Davis *et al.*, 1999; Wippersteg *et al.*, 2002; Heyers *et al.*, 2003). Isolation of endogenous forms of *mariner*-like elements from schistosomes, if indeed they are present in the schistosome genome, might facilitate development of target species-specific constructs. (*mariner*-like elements occur in planarians [Garcia-Fernàndez *et al.*, 1995].)

Retroviruses such as Moloney murine leukemia virus (Mo-MLV) (Fig. 2-2) have been widely used for transgenesis studies, including human gene therapy research. Although retroviruses exhibit narrow host ranges, pseudotyping Mo-MLV with the envelope glycoprotein (G) of the vesicular stomatitis virus (VSV) expands its host range (Yee *et al.*, 1994). It is also feasible that Mo-MLV constructs pseudotyped with envelope proteins from endogenous retroviruses, for example the envelope of *Tas* from *A. lumbricoides* (Fig. 2-2), would enhance the parasite species cell-specificity of these constructs, and circumvent host cell-specific entry blocks in *Ascaris* and related nematodes. This approach appears feasible for schistosomes once envelope genes from endogenous retrovirus-like retrotransposons from schistosomes are characterized. A desirable transgenesis vector would be one that already integrates into the schistosome genome as part of the vector's natural life cycle. The active search for and characterization of endogenous MGEs in schistosomes have already yielded several possible candidates (Figs. 2-1-2-4). It seems feasible that schistosome retrotransposons such as *Boudicca* and *Gulliver* can be adapted directly as transgenesis vectors given similar success with other mobile elements (Garraway *et al.*, 1997; Ivics *et al.*, 1997; Dewannieux *et al.*, 2003). We are pursuing this and similar lines of investigation in our laboratories, towards the goal of development of a tractable system for schistosome transgenesis.

ACKNOWLEDGEMENTS

PJB is a recipient of a Burroughs Wellcome Fund Scholar Award in Molecular Parasitology. Findings described here were supported in part by a grant from the Deutsche Forschungsgemeinschaft (KA 866/2-1) to BHK.

REFERENCES

Bae, Y.A., Moon, S.Y., Kong, Y., Cho, S.Y., and Rhyu, M.G., 2001, *CsRn1*, a novel active retrotransposon in a parasitic trematode, *Clonorchis sinensis*, discloses a new phylogenetic clade of *Ty3/gypsy*-like LTR retrotransposons, *Mol. Biol. Evol.* **18:**1474-1483.

Boyle, J.P., and Yoshino, T.P., 2003, Gene manipulation in parasitic helminths. *Int. J. Parasitol.* **33:**1259-1268.

Brindley, P.J., Laha, T., McManus, D.P., and Loukas, A., 2003, Mobile genetic elements colonizing the genomes of metazoan parasites, *Trends Parasitol.* **19:**79-87.

Charlesworth, B., Sniegowski, P., and Stephan, W., 1994, The evolutionary dynamics of repetitive DNA in eukaryotes, *Nature* **371:**215-220.

Copeland, C.S., Brindley, P.J., Heyers, O., Michael, S.F., Johnston, D.A., Williams, D.J., Ivens, A., and Kalinna, B.H., 2003, *Boudicca*, a retrovirus-like, LTR retrotransposon from the genome of the human blood fluke, *Schistosoma mansoni*, *J. Virol.* **77**:6153-6166.

Davis, R.E., Parra, A., LoVerde, P.T., Ribeiro, E., Glorioso, G., and Hodgson, S., 1999, Transient expression of DNA and RNA in parasitic helminths by using particle bombardment. *Proc. Natl. Acad. Sci. USA* **96**:8687-8692.

Deininger, P.L., and Batzer, M.A., 2002, Mammalian retroelements, *Genome Res.* **12**:1455-1465.

Dewannieux, M., Esnault, C., and Heidmann, T., 2003, LINE-mediated retrotransposition of marked Alu sequences, *Nat. Genet.* **35**:41-48.

Drew, A.C., and Brindley, P.J., 1995, Female-specific sequences isolated from *Schistosoma mansoni* by representational difference analysis, *Mol. Biochem. Parasitol.* **71**:173-181.

Drew, A.C., and Brindley, P.J., 1997, A retrotransposon of the non-long terminal repeat class from the human blood fluke *Schistosoma mansoni*. Similarities with the chicken repeat 1-like elements from vertebrates, *Mol. Biol. Evol.* **14**:602-610.

Drew, A.C., Minchella, D.J., King, L.T., Rollinson, D., and Brindley, P.J., 1999, *SR2*, non-long terminal repeat retrotransposons of the RTE-1 lineage, from the human blood fluke *Schistosoma mansoni*, *Mol. Biol. Evol.* **16**:1256-1269.

Fan, J., Minchella, D.J., Day, S.R., McManus, D.P., Tiu, W.U., and Brindley, P.J., 1998, Generation, identification, and evaluation of expressed sequence tags from different developmental stages of the Asian blood fluke *Schistosoma japonicum*, *Biochem. Biophys. Res. Commun.* **252**:348-356.

Fan, J., and Brindley, P.J., 1998, Retrotransposable elements in the *Schistosoma japonicum* genome, in: *Proceedings of the 9th International Congress of Parasitology (ICOPA IX)*, I. Tada, S. Kojima, and M. Tsuji, eds., Monduzzi Editore, Bologna, Italy, pp. 821-825.

Ferbeyre, G., Smith, J.M., and Cedergren, R., 1998, Schistosome satellite DNA encodes active hammerhead ribozymes, *Mol. Cell. Biol.* **18**:3880-3888.

Finnegan, D.J., 1992, Transposable elements, *Curr. Opin. Genet. Dev.* **2**:861-867.

Foulk, B.W., Pappas, G, Hirai, H, and Williams, D.L., 2002, Adenylsuccinate lyase of *Schistosoma mansoni*: gene structure, mRNA expression, and analysis of the predicted peptide structure of a potential chemotherapeutic target, *Int. J. Parasitol.* **32**:1487-1495.

Garcia-Fernàndez, J., Bayascas-Ramírez, J-R., Marfany, G., Muñoz-Mármol, A.M., Casali, A., Baguñà, J., and Saló, E., 1995, High copy number of highly similar *mariner*-like transposons in planarian (Platyhelminthe): evidence for a trans-phyla horizontal transfer, *Mol. Biol. Evol.* **12**:421-431.

Garraway, L.A., Tosi, L.R., Wang, Y., Moore, J.B., Dobson, D.E., and Beverley, S.M., 1997, Insertional mutagenesis by a modified in vitro *Ty1* transposition system, *Gene* **198**:27-35.

Gueiros-Filho, F.J., and Beverley, S.M., 1997, Trans-kingdom transposition of the *Drosophila* element *mariner* within the protozoan *Leishmania*, *Science* **276**:1716-1719.

Heyers, O., Walduck, A.K., Brindley, P.J., Bleiß, W., Lucius, R., Dorbic, T., Wiittig, B., and Kalinna, B.H., 2003, *Schistosoma mansoni* miracidia transformed by particle bombardment infect *Biomphalaria glabrata* snails and develop into transgenic sporocysts, *Exp. Parasitol.* **105**:174-178.

Hirai, H., Taguchi, T., Saitoh, Y., Kawanaka, M., Sugiyama, H., Habe, S., Okamoto, M., Hirata, M., Shimada, M., Tiu, W.U., Lai, K., Upatham, E.S., and Agatsuma, T., 2000, Chromosomal differentiation of the *Schistosoma japonicum* complex, *Int. J. Parasitol.* **30**:441-452.

Holt, R.A., Subramanian, G.M., Halpern, et al., 2002, The genome sequence of the malaria mosquito *Anopheles gambiae*, *Science* **298**:129-149.

Hu, W., Yan, Q., Shen, D-K., Liu, F., Xu, X-R., Zhu, Z-D., Wu, X-W., Zhang, X., Wang, J-J., Xu, X., Wang, Z., Huang, J., Wang, S-Y., Wang, Z-Q., Brindley, P.J., McManus, D.P., Xue, C-L., Feng, F., Chen, Z., and Han, Z-G., 2003, Evolutionary and biomedical implications of a *Schistosoma japonicum* complementary DNA resource, *Nat. Genet.* **35**:139-147.

Ivics, Z., Hackett, P.B., Plasterk, R.H., and Izsvak, Z., 1997, Molecular reconstruction of *Sleeping Beauty*, a *Tc1*-like transposon from fish, and its transposition in human cells, *Cell* **91**:501-510.

Kajikawa, M., and Okada, N., 2002, LINEs mobilize SINEs in the eel through shared 3' sequence, *Cell* **111**:433-444.

Kazazian, H.H., 1999, An estimated frequency of endogenous insertional mutations in human. *Nat. Genet.* **22**:130.

Laha, T., McManus, D.P., Loukas, A., and Brindley, P.J., 2000, *Sjα* elements, short interspersed element-like retroposons bearing a hammerhead ribozyme motif from the genome of the Oriental blood fluke *Schistosoma japonicum, Biochim. Biophys. Acta* **1492**:477-482.

Laha, T., Brindley, P.J., Smout, M.J., Verity, C.K., McManus, D.P., and Loukas, A., 2002a, Reverse transcriptase activity and UTR sharing of a new *RTE*-like, non-LTR retrotransposon from the human blood fluke, *Schistosoma japonicum, Int. J. Parasitol.* **32**:1163-1174.

Laha, T., Brindley, P.J., Verity, C.K., McManus, D.P., and Loukas, A., 2002b, *pido*, a non-long terminal repeat retrotransposon of the chicken repeat 1 family from the genome of the Oriental blood fluke, *Schistosoma japonicum, Gene* **284**:149-159.

Lander, E.S., Linton, L.M., Birren, B., et al., 2001, Initial sequencing and analysis of the human genome, *Nature* **409**:860-921.

Malik, H.S., Burke, W.D., and Eickbush, T.H., 1999, The age and evolution of non-LTR retrotransposons, *Mol. Biol. Evol.* **16**:793-805.

Malik, H.S., Henikoff, S., and Eickbush, T.H., 2000, Poised for contagion: evolutionary origins of the infectious abilities of invertebrate retroviruses, *Genome Res.* **10**:1307-1318.

O'Brochta, D.A., Sethuraman, N., Wilson, R., Hice, R.H., Pinkerton, A.C., Levesque, C.S., Bideshi, D.K., Jasinskiene, N., Coates, C.J., James, A.A., Lehane, M.J. and Atkinson, P.W., 2003, Gene vector and transposable element behavior in mosquitoes, *J. Exp. Biol.* **206**:3823-3834.

Simpson, A.J.G., Sher, A., and McCutchan, T., 1982, The genome of *Schistosoma mansoni*: isolation of DNA, its size, bases and repetitive sequences, *Mol. Biochem. Parasitol.* **6**:125-137.

Spotila, L.D., Hirai, H., Rekosh, D.M., and LoVerde, P.T., 1989, A retroposon-like short repetitive DNA element in the genome of the human blood fluke, *Schistosoma mansoni, Chromosoma* **97**:421-428.

Verjovski-Almeida, S., DeMarco, R., Martins, E.A.L., Guimarães, P.E.M., Ojopi, E.P.B., Paquola, A.M., Piazza, J.P., Nishiyama, M.Y. Jr, Kitajima, J.P., Adamson, R.E., Ashton, P.D., Bonaldo, M.F., Coulson, P.S., Dillon, G.P., Farias, L.P., Gregorio, S.P., Ho, P.L., Leite, R.A., Malaquias, L.C.C., Marques, R.C.P., Miyasato, P.A., Nascimento, A.L.T.O., Ohlweiler, F.P., Reis, E.M., Ribeiro, M.A., Sá, R.G., Stukart, G.C., Soares, M.B., Gargioni, C., Kawano, T., Rodrigues, V., Madeira, A.M.B.N., Wilson, R.A., Menck, C.F.M., Setubal, J.C., Leite, L.C.C., and Dias-Neto, E., 2003, Transcriptome analysis of the acoelomate human parasite *Schistosoma mansoni, Nat. Genet.* **35**:148-157.

Wippersteg, V., Kapp, K., Kunz, W., Jackstadt, W.P., Zahner, H., and Grevelding, C.G., 2002, HSP70-controlled GFP expression in transiently transformed schistosomes, *Mol. Biochem. Parasitol.* **120:**141-150.

Yee, J.K., Friedmann, T., and Burns, J.C., 1994, Generation of high-titer pseudotyped retroviral vectors with very broad host range, *Methods Cell. Biol.* **43:**99-112.

Chapter 3

GENDER-SPECIFIC BIOLOGY OF SCHISTOSOMA MANSONI:
Male/female interactions

Philip T. LoVerde*, Edward G Niles, Ahmed Osman and Wenjie Wu
*Department of Microbiology and Immunology, School of Medicine and Biomedical Sciences, State University of New York, Buffalo, New York 14214. *Corresponding author: loverde@buffalo.edu*

Key words: Schistosomiasis, host-parasite interactions, signaling pathways

1.   INTRODUCTION

Eggs produced by worm pairs are important in transmission of the parasite and responsible for pathogenesis. *Schistosoma mansoni* worm pairs produce approximately 300 eggs per day. Approximately half of the deposited eggs reach the outside environment to continue the life cycle. The other 50% are swept into the circulation and filter out in the periportal tracts of the liver eliciting granulomatous inflammatory reactions that can lead to periportal fibrosis, portal hypertension and the serious sequelae of intestinal schistosomiasis such as hepatosplenomegaly and esophageal and gastric varicies (1).   Thus, understanding the molecular basis for male-female interactions that lead to female reproductive development should offer targets to prevent egg production and thus prevent both transmission of the parasite and morbidity due to the eggs.

2.   MALE/FEMALE DEVELOPMENT

S. mansoni has 8 pairs of chromosomes.   The female schistosome is ZW and the male is ZZ (2). Sex is determined by a chromosomal mechanism in the zygote that develops into the miracidium. The miracidium then transfers

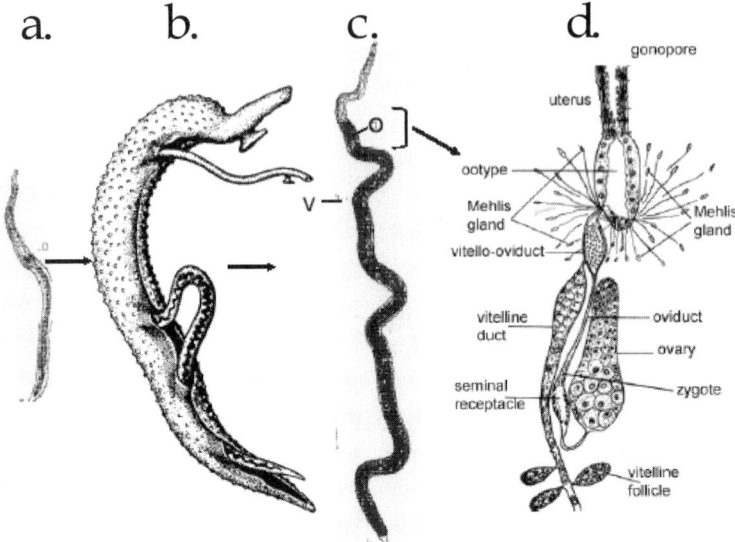

Figure 3-1. Female *Schistosoma mansoni,* showing (a) an immature female from a single-sex infection, (b) male mated with a female, (c) a mature female from a bisexual infection, and (d) enlargement of the mature female reproductive system. Oocytes produced in the ovary are released into an oviduct. A dilitated region in the oviduct, the seminal receptacle, is the site of fertilization. The fertilized egg moves down the oviduct to a region where it meets the vitelline duct, the vitello-oviduct. Here, approximately 38 vitelline cells surround each fertilized egg. This bolus of material moves into the ootype that is filled with mucus secretions presumably from the Mehlis' gland. With the contraction of the ootype, which determines the shape of the egg, granules are released from the surrounding vitelline cells. The materials from the granules (vitelline droplets) begin to coalesce (crosslink), presumably due to the action of a phenol oxidase enzyme(s) and the eggshell begins to form. The egg is released into the uterus and eventually deposited by the female worm through the gonopore.

the germline from the human host to the snail host. Inside the snail the parasite undergoes asexual reproduction producing thousands of cercariae. Thus, a single miracidium will give rise to a clonal population of cercariae that are all of the same sex. Male schistosomes undergo normal morphological development in the mammalian host whether isolated from a single-sex or bisexual infection. However, behavioral, physiological and antigenic differences between males from single, as opposed to bisexual, infections have been reported (reviewed in 3 and 4). Female schistosomes from single sex infections show distinct differences from females obtained from bisexual infections (3-6, Fig. 3-1). Females from single sex infections of the mammalian host are underdeveloped in that they are stunted and exhibit an immature reproductive system. The ovary, ootype and uterus are developed in the immature female; but the vitellaria, whose cells produce the eggshell precursors and nutrients for the egg, and the Mehlis' gland, which surrounds the ootype where eggs are formed, are not fully developed.

3. THE MALE STIMULATES FEMALE GROWTH AND REPRODUCTIVE DEVELOPMENT

Female worm physical and reproductive maturation is dependent on the presence of mature male worms (3-6, Figs. 3-1, 3-2). Female worms are not stimulated to mature until the males themselves have become mature. The stimuli for female growth and for reproductive development appear to be independent (7). The male stimulus is independent of sperm transfer and fertilization, is not species- or genus-specific, and is independent of central nervous system control (3-7). Whether the male stimulus responsible for female sexual maturation involves a physical, tactile relationship that may involve nutrition, a chemical transfer in which the male provides some hormone, nutrient or messenger to the female or a combination of thigmotactic and chemotactic stimuli has yet to be determined (3-10). However, DNA synthesis dramatically increases when females from single-sex infections are paired with male worms indicating mitotic activity (11, 12). An intimate association between the male and female worm, which is achieved by the female residing within a ventral groove, the gynaecophoric canal of the male, is necessary and sufficient to direct female growth and reproductive development (Fig. 3-2). Thus the male stimulus that requires contact is spatially distributed throughout the gynaecophoric canal. However, its effect (vitelline gland development) is local, that is, where contact occurs, and it is not longitudinally propagated along the female worm. The stimulus of the male worm is not only necessary for female worms to initiate and complete physical and reproductive development but also for the female to maintain her mature state. Studies (13) have shown that egg-laying female schistosomes (containing stored sperm) but without male partners, surgically implanted into naive animals cease laying eggs and begin to regress physically and reproductively to the immature female state. Adult worm pairs, similarly transplanted, continue to produce viable eggs. When male worms are introduced, months later, with the widowed female schistosomes, pairing occurs and normal reproductive activity resumes. Molecular studies also have demonstrated that female-specific gene expression is dependent on pairing with male worms (14, 15). Not surprisingly, the female schistosome has a stimulatory effect on the male as evidenced by changes in levels of glutathione and lipids in the male and expression of new antigens in the gynecophoric canal in the male, all in response to mating with female worms (16, 17).

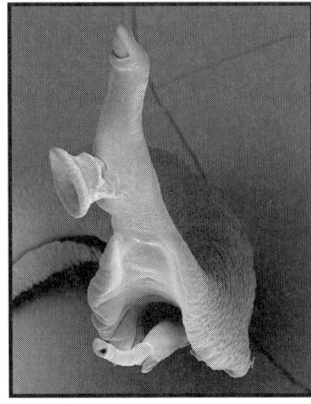

Figure 3-2. Mature adult *Schistosoma mansoni* worms encopula. The long slender female resides in the gynecophoric canal of the male (Source David Scharf).

4.   OTHER ROLES FOR MALE IN FEMALE DEVELOPMENT

The male sex has evolved a number of roles in the life history of the female (7, 18, 19). Muscularity of the male compared to the female is one example (Figs. 3-1, 3-2). The male plays a critical role in transporting the female who by herself is not capable of migrating against the flow of blood from the portal sites in the liver to the smaller venules and capillaries where her eggs can be deposited to exit the vertebrate host. The muscularity of the male schistosome allows them to assist the weaker females to physically pump host blood into her intestine by a massaging action of the muscular walls of the gynecophoric canal. The male also plays a role in passage of metabolites such as glucose and cholesterol and glycoproteins from the paired male to the gynecophoric canal-residing female worm (e.g. 20-22). This has lead to the notion that the pairing with the male worm initiates a change in the tegumental uptake characteristics of the female that facilitates her development and maintenance.

5.   MOLECULAR BASIS FOR MALE-FEMALE INTERACTIONS

Studies were initiated to identify the underlying molecular mechanisms for the classical observation that female schistosomes do not complete reproductive development unless a male worm is present. Initially, a number of genes expressed only in reproductively active female worms were identified (3, 14, 23-25). The best-characterized genes are the p14 and p48 eggshell protein precursor (chorion) genes of *S. mansoni*. These genes, like the others, show sex-, temporal- and tissue-specific expression during female development (3, 4, 14, 15). The messenger RNAs (mRNA) encoding p14 and p48 are present only in mature female worms where they are first detected at the time of worm pairing and increase to a high level with the start of egg production. p14 and p48 mRNA are synthesized and translated in vitelline cells and the gene products are associated with the vitelline droplets (the proteinacious granules that contain the eggshell precursor proteins) (14). Thus the male stimulus is associated with pairing and results in the regulation of vitelline cell (female-specific) gene expression (Fig. 3-1).

6. EGG PRODUCTION

Each mature *S. mansoni* female worm with an average life span of five years produces 300 eggs per day (7). In order to maintain that level of reproductive activity, the worm pairs continuously synthesize glycogen (26), the male transfers glucose to the female (20) and the female worm, as has been estimated, converts nearly her own body weight into eggs every day (27). The high demands of egg production dictate an active metabolic role for the vitelline gland. In order to maintain an output of one egg every 4.8 min, each female worm produces more than 11,000 mature vitelline cells per day (5, Fig. 3-1).

In schistosomes, choriogenesis is continuous (but male dependent) and, involves at least two different genes (p14 and p48) that are each represented by a few (five) copies on chromosome 3 (14, 28, 29). There is no evidence for gene amplification or rearrangement of eggshell protein genes during schistosome development (3) as has been reported for Drosophila and silkmoth choriogenesis. Schistosomes meet such a high rate of egg production by the continuous synthesis of a large number of vitelline cells and a high amount of p14 and p48 transcripts (42×10^8 and 2.8×10^8 molecules per μg of total RNA, respectively, 14) in each cell. Other female-specific genes have been identified (30, 24, 23, 25).

Using a cDNA microarray, Hoffman et al. (31) identified 12 female-specific and 4 male-specific novel transcripts. One gene was a putative tyrosinase gene that showed female-specific expression. This gene product is thought to play a critical role in eggshell formation (32). As the sequence of the *S. mansoni* genome becomes available, we will be in a position to learn more about those genes that respond to a male stimulus (Williams and Pierce, Chapter 4; 33).

7.   CO-ORDINATE GENE EXPRESSION

As the stage, tissue and temporal patterns of gene expression are the same for the p14 and p48 genes, one might expect the same *trans*-acting factors to bind to conserved *cis*-acting motifs (DNA sequences) in order to co-ordinately regulate transcription of a number of genes involved in female reproductive development in response to the male stimulus.

Indeed, for those genes where the 5' end of the female-specific mRNA has been defined and the upstream regions characterized, a number of putative *cis*-acting DNA elements have been identified (3, 14).   As *S. mansoni* eggshell genes are like the chorion genes of Drosophila and silkmoth in that they are developmentally regulated, showing sex-, tissue- and temporal-specific expression, it was interesting that motifs were identified in schistosome upstream sequences similar to those found in silkmoth and Drosophila (3). A Drosophila transcription factor (Ultraspiracle) that binds to a region containing a TCACGT core sequence was identified as a member of the RXR family of steroid hormone receptors (34). A similar sequence is present as part of an imperfect palindrome in the upstream region of p14. Furthermore a direct repeat that follows the "3-4-5" rule of steroid response elements is present in the p14 upstream region (4, 35). This later sequence consists of two identical half-sites, AACTATCA, spaced by 5 nucleotides. Presence of these *cis*-elements led to a search for schistosome homologues of nuclear receptors (35).

8.   ROLE OF NUCLEAR RECEPTORS IN REGULATING SCHISTOSOME GENE EXPRESSION

Nuclear receptors (NR) play an essential role in the regulation of cell growth and differentiation by providing a direct link between signaling molecules and the transcription response. NR comprise a large, complex superfamily of structurally related transcription regulatory factors in metazoans (36, 37).

Nested PCR permitted the identification of segments of the DNA binding domain of five nuclear receptor homologues from *Schistosoma mansoni,* exhibiting sequence relatedness to the vertebrate RXR, TR4, FTZ-F1, COUP I and COUP II receptors (38).   Independently, the first full length nuclear receptor cDNA from *S. mansoni* was cloned, which encoded the RXR 1 receptor homologue (35).   Subsequently, a full length cDNA for a second RXR homologue, RXR 2, was identified that exhibited little sequence conservation with RXR 1 outside of the DNA binding region (39, 40). Both genes were transcribed in all developmental stages. *In situ* hybridizations showed that mRNA for both proteins are present in several adult worm cells which supports involvement of RXR 1 and RXR 2 in a variety of adult functions. Importantly, both RXR 1 and RXR 2 mRNA were found in the vitellaria in mature females. This latter observation is particularly important as each protein was proposed to play a role in the sex specific expression of the chorion p14 gene in vitelline cells.

Both RXR 1 and RXR 2 were shown to stimulate gene expression in yeast when multiples of the p14 gene upstream region were placed upstream of the promoter in a yeast reporter plasmid (39, 41). Further studies showed that RXR 1 interacted with a direct repeat sequence in the p14 gene promoter, both *in vitro* and in yeast (35). The RXR 2 binding is more promiscuous and the target sequence remains undefined. It is clear that RXR 1 and RXR 2 do not functionally interact *in vitro* or in yeast (41). A comparison of the two protein sequences identify other interesting motifs that suggest RXR 1 and RXR 2 carry on very different functions *in vivo*. For example, RXR 1 has a sequence motif present in other ligand dependent nuclear receptors that is part of the co-activator interaction site. This sequence is not present in RXR 2, which raises the possibility that RXR 1 but not RXR 2 is ligand dependent.

RXR nuclear receptors are active as heterodimers in all other systems. One can anticipate that the functional properties of the individual RXR subunits will be extensively modified by the inclusion of an additional subunit. Recently, two novel nuclear receptors were identified in *S. mansoni* that are related in sequence to the vertebrate CAR (constitutive androstane receptor) and a retinoid receptor (unpublished data). Both nuclear receptors were shown to interact with RXR1 but not with RXR2. These results highlight the potential functional differences between *S. mansoni* RXR1 and RXR2 and provide the first steps in determining the functional role of these nuclear receptors in schistosome biology.

Additional nuclear receptors have been identified in *S. mansoni* such as FTZ-F1 (42). *In vitro*, smFTZ-F1 binds to DNA targets as a monomer with the same specificity as SF-1, the FTZ-F1 homologue of higher eukaryotes. Recently, a closely related cDNA was cloned indicating that smFTZ-F1 is a member of a small family of nuclear receptors (unpublished data). Based on information derived from other eukaryotes, one can anticipate that smFTZ-F1 will play essential roles in schistosome development and differentiation.

Studies on *S. mansoni* nuclear receptors have just begun. Through the application of nested PCR, library screening, and database searching, additional putative nuclear receptors have been identified and their roles in schistosome biology are under investigation. With the anticipation of the publication of the results of large-scale cDNA sequencing projects and an initial genomic DNA sequencing venture, a complete repertoire of nuclear receptor genes will be defined (Williams and Pierce, Chapter 4). Several fundamental questions such as the number of nuclear receptors present in schistosomes, which are ligand regulated and which are true orphans, are the ligands of parasite and/or host origin, and can anti-helmenthic drugs be synthesized which selectively block individual schistosome nuclear receptors, will be able to be addressed.

9.   SIGNALING PATHWAYS

Signaling pathways must be employed for a male to stimulate and maintain female vitelline cell development. Evidence for a role for signal transduction in female reproductive development comes from a number of studies. For example the tyrosine receptor kinase coupled pathway in which closely related proteins belonging to the superfamily of small GTPases act as molecular switches, which relay a signal transmitted from receptor tyrosine-kinases (RTK), activated by a wide variety of growth factors and peptide hormones, through a phosphorylation cascade of downstream effectors that ends in the nucleus where a specific outcome is orchestrated.

Ras expression is regulated in male and female parasites suggesting that this pathway functions in male-female interactions (43, 44). It is hypothesized that the male signal regulates the multiplication (mitosis) and differentiation of germ cells that are already present in the virgin female to form the vitellaria (4, 6). As the Ras pathway is known to regulate cell proliferation events in other organisms, its role in transducing a signal for mitogenic proliferation in female reproductive development is hypothesized (6, 45). Other examples of *S. mansoni* factors that likely have a role in male and/or female growth, differentiation and development are Rho that also seems preferentially expressed in adult female worms (46), receptor tyrosine kinases (RTK) SmRTK-1 and SmRTK-2 that share characteristic features with insulin receptors (47) and a Syk-family tyrosine kinase (SmTK-4) that is preferentially transcribed in reproductive organs of both sexes (48).

The TGFβ (serine/threonine receptor kinase coupled pathway) signaling pathway has been extensively studied in our laboratory (Fig. 3-3). It comprises a family of structurally related polypeptide growth factors, each capable of regulating an array of cellular processes (49). TGFβ and related factors signal through transmembrane protein serine/threonine kinases (Fig. 3-3). There are two types of surface exposed receptor ser/thr kinases: type I (RI) and type II (RII). Binding of a ligand to a type II receptor in concert with a type I receptor leads to the formation of an activated receptor complex. Following ligand binding, the type II receptor kinase phosphorylates and thereby activates the corresponding type I receptor. Activated RI then transduces the signal to a family of Smad proteins that carry the signal to the nucleus and regulate transcription of selected genes in response to ligand. Smads function as intracellular signaling effectors for the TGFβ family (50). Members of the Smad family that receive a signal from the activated receptor complex are called receptor-regulated Smads (R-Smads). R-Smads demonstrate specific receptor activation. Smad-2 and -3 are phosphorylated and activated by the activin and TGFβ receptors; Smads-1, -5, and -8 are activated by the BMP receptors. Phosphorylated, R-Smad forms a complex with a common member of the Smad family, Smad-4 and the heterooligomer is translocated into the nucleus. The heteromeric complex binds to other regulatory proteins in the nucleus that direct the Smad complex to target gene promoters to regulate specific gene expression. Some Smad family members (Smad-6 and -7) act as suppressors and block the signal to the nucleus.

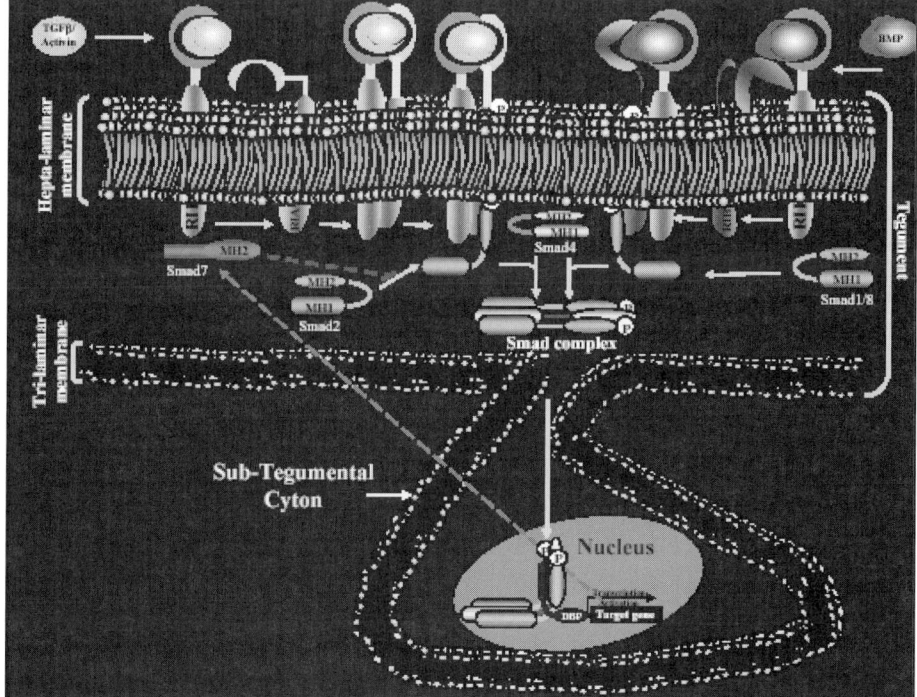

Figure 3-3. Hypothetical TGFβ signaling pathway in *Schistosoma mansoni* showing the TBFβ /Activin and the Bone morphogenic (BMP) pathways. The identified schistosome pathway members TbRII, TbRI, SmSmads 1, 2, 3, 8/9, and 4 are identified along with a hypothetical Smad7. See text for details.

To date, five members of the schistosome Smad family have been isolated: *Sm*Smad-1 (51), *Sm*Smad-2 (51, 52), *Sm*Smad-3 (unpublished), *Sm*Smad-4 (53), and *Sm*Smad 8/9 (54) and two TGFβ-like receptors, SmTβRI, (55) and SmTβRII, (54) (Fig. 3-3). Interestingly, SmTβRI was shown to be expressed at the parasite surface and this expression was up-regulated after infection of the mammalian host (55). Importantly, *Sm*Smad-2 and SmSmad-4 localize to the vitelline cells, among other cell types, as demonstrated by immunolocalization and *in situ* hybridization experiments (53). In functional assays, *Sm*Smad-2 is phosphorylated by the activated form of *Sm*TβRI and translocates to the nuclei of Mink lung epithelial cells upon treatment with the ligand TGFβ (52). These data and other results (56) indicate that *Sm*Smad-2 responds to TGFβ signals perhaps of host origin by interaction with the surface exposed TGFβ receptor I, which phosphorylates it, and in cooperation with *Sm*Smad-4 (53), translocates to the nucleus presumably to regulate target gene transcription and consequently elicit a specific TGFβ effect. This has yet to be demonstrated in the schistosome worm itself. However, such studies of signaling open a new era for investigation of parasite-parasite interactions.

As presented, this simple linear pathway of TGFβ signaling is a minor component of a very complex signaling network, where crosstalk, feedback loops, branch points and multicomponent signaling complexes compose the current view of TGFβ -mediated (and RAS-mediated for that matter) signaling pathways.

10.   ROLE OF HOST CYTOKINES IN REPRODUCTIVE DEVELOPMENT

Amiri et al. (58) reported that the host cytokine, tumor necrosis factor (TNF) α works as a positive regulator of fecundity for female parasites both *in vivo* and *in vitro*. However, others have been unable to confirm these results (59). More recently, Davies et al. (60) identified a set of signals from a novel population of hepatic CD4+ T cells that were able to modulate schistosome parasite development to yield a stunted phenotype. Using a series of KO mice, the authors were able to demonstrate that a CD4+ T lymphocyte population that share a similar tissue distribution to the developing schistosome are important in sending an as yet unknown signal that regulates development. The authors suggest that in immunodeficient hosts, where the schistosome fails to receive appropriate signals from the immune system, schistosome development is altered such that attenuated (arrested development) forms appear that prolong the survival of the parasite and host until optimal conditions for growth, mating, fecundity and transmission return. The next obvious question to address is the identification and nature of the products of the hepatic CD4+ T cells that modulate parasite development.

IL-7 was shown to be an important cytokine in regulating schistosome development. It was identified as being important in the initial phases of *S. mansoni* infection by impairing parasite migration to the lungs, increasing the number of surviving adult worms, and resulting in more severe liver pathology (61). Interestingly, in the absence of IL-7 female worms showed altered fecundity, resulting in decreased egg counts, and increased non-viable eggs trapped in the tissues and an amelioration of pathology. IL-7 has a dramatic effect on adult worm development. In its absence, adult male and female worms are fully developed yet stunted in size. However, the lack of direct binding of radiolabeled IL-7 to the parasite surface suggested that IL-7 acts indirectly (61, 62). As the effect of IL-7 deficiency on parasite growth, that is, dwarfism, was similar to that described in studies on hypothyroid mice (63), Saule et al (64) addressed the question of whether there was an interaction between IL-7 and thyroid hormones in schistosome host-parasite interactions. Treatment of mice with thyroxine (T4), the main form of thyroid hormone synthesized in the thyroid gland, resulted in higher worm burdens, larger (giant) worms, an increase in the number of hepatic eggs but no increase in fecundity or change in hepatic collagen deposition compared to control mice. In contrast mice that were made hypothyroid by being feed an iodine-deficient diet, before and during infection, the number of worms recovered was similar to control animals, however a larger number of immature worms were present which resulted in reduced fecundity and liver collagen deposition. Amazingly, co-administration of T4 and IL-7 to mice had a dramatic effect on worm growth producing "super giant" worms that exhibited an overall proportional increase in size (64). In spite of the increase in size the number of hepatic eggs and hepatic pathology remained similar to control mice.

Thus the early cutaneous events occurring after *S. mansoni* penetration result in IL-7 production that has a profound effect on infection outcome including male and female growth and development. IL-7 seems to have a significant role throughout development including male and female growth.

It is obvious from the above mentioned studies that the host is not a passive container providing nutrients for the parasite but rather that the host provides essential factors for schistosome growth, reproductive development, and egg production.

11.   A MODEL FOR MALE-FEMALE INTERACTIONS

In order for the male to stimulate and to maintain female vitelline cell development, there must be direct contact between the male and the female (Figs. 3-1, 3-2). This is accomplished by the female residing in the gynecophoric canal of the male. Whether the male signal is thigmotactic or chemotactic is unknown. However, as the worms live in the circulation even the transfer of a chemical signal requires close contact. If one considers the architecture of the female parasite, it seems that the transduction of the signal likely involves a number of intercellular as well as intracellular substrates to transduce the signal from the surface of the male to the surface of the female to the target primordial vitelline cells and subsequent vitelline cell maturation.

The surface of the female worm is composed of a tegument that is in direct contact with the host environment, and the male parasite. The tegument overlays circular and longitudinal muscle bundles beneath which are found the organ systems including the vitellaria in mature parasites, or vitelline primordia in immature parasites (65). There are several scenarios for male signaling. The signal transferred directly from the male could be by a male-derived ligand that binds to a receptor on the surface (tegument) of the female (Fig. 3-4, stage 1). The ligand could be released or be contact-dependent (mechanical association) via membrane bound molecules. Alternatively, the male may release a small hydrophobic signaling ligand that diffuses across the female apical heptalaminar membrane and binds on the cytoplasmic side to a cytoplasmic face receptor (CR, Fig 3-4, stage 2) in the tegument. Also, the female residing in the gynecophoric canal could first stimulate the male to produce a signal that is then sent to the female by one of the above scenarios (Fig 3-4, stage 3 and insert). In any case the ligand binds the receptor and tranduces a signal via one or more pathways (eg, Ras and/or TGFβ signaling pathways) from the tegument to the target vitelline cell receptor which may be on the vitelline cell membrane and through intercellular substrates to the vitelline cell nucleus where vitelline cell gene expression is activated (see Fig. 3-4, stages 4 and 5). The activation must stimulate proliferation for the vitellaria to form from the primordial cells and subsequently must direct differentiation for each vitelline cell to undergo development from an immature stage 1 (S1) cell to a mature stage 4 (S4, 66) that will participate in egg production. In addition other cells that contribute directly or indirectly to reproductive development and to growth of the female likely to receive signals. Some of the signals as the evidence suggests, are separate and others may be the same but involve "cross talk" between different signaling pathways.

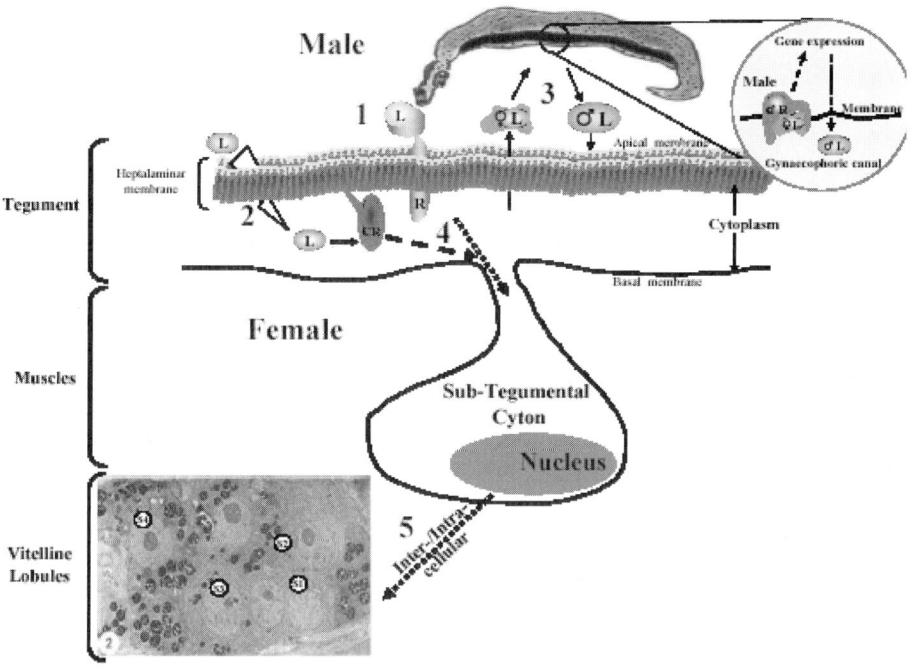

Figure 3-4. Model depicting various scenarios of the mechanism that the male schistosome employs to regulate female gene expression and subsequent reproductive development. See text for detailed description (modified from LoVerde, 2002). L, ligand; •L, female L; •L, male L; R, receptor; •R, male R; CR, cytoplasmic R; S1-S4, stage 1-4. Insert shows region of male gynaecophric canal.

The data available to date support the notion that nuclear receptors (*Sm*RXR's) and signaling pathways (Ras and TGFβ) play roles in female-specific gene regulation by localizing to vitelline cells and in the case of *Sm*RXR's binding to *cis*-elements present in the upstream region of the *p14*. Having identified nuclear transcription factors that are at the terminal end of male-stimulated female gene expression, one can move upstream in the pathway to identify the relevant signaling pathways employed and eventually define the nature of the male stimulus and the female receptor.

In the schistosomes an interesting biological interplay has evolved such that male schistosomes via an unknown stimulus regulate female-specific gene expression and thus female reproductive development and egg production. As the egg stage is responsible for the pathogenesis in schistosomiasis, understanding the molecular basis for this interesting biological interplay will lead to the identification of novel targets for drug and vaccine intervention. Signaling pathways and nuclear receptors are prime candidates for transduction of the male stimulus to the vitelline cells of the female parasite.

ACKNOWLEDGMENTS

The authors thank David Scharf for Figure 3-1. This research is supported in part by NIH grant AI46762.

REFERENCES

1. Hoffmann K.F., Wynn T.A., Dunne D.W. Cytokine-mediated host responses during schistosome infections; walking the fine line between immunological control and immunopathology. Adv Parasitol 2002;52:265-307.
2. Short R.B. Presidential address: Sex and the single schistosome. J Parasitol 1983;69:3-22.
3. LoVerde P.T., Chen L. Schistosome female reproductive development. Parasitology Today 1991;7:303-308.
4. LoVerde P.T. Presidential address. Sex and schistosomes: an interesting biological interplay with control implications. J Parasitol 2002;88:3-13.
5. Popiel I. The reproductive biology of schistosomes. Parasitology Today 1986;2:10-15.
6. Kunz W. Schistosome male-female interaction: induction of germ-cell differentiation. Trends Parasitol 2001;17:227-31.
7. Basch P.F. Schistosomes, Development, Reproduction, and Host Relations. New York, Oxford: Oxford University Press, 1991.
8. Ribeiro-Paes J.T., Rodrigues V. Sex determination and female reproductive development in the genus *Schistosoma*: a review. Rev Inst Med Trop Sao Paulo 1997;39:337-44.
9. de Mendonca R.L., Escriva H., Bouton D., Laudet V., Pierce R.J. Hormones and nuclear receptors in schistosome development. Parasitol Today 2000;16:233-40.
10. Salzet M., Capron A., Stefano G.B. Molecular crosstalk in host-parasite relationships: schistosome- and leech-host interactions. Parasitol Today 2000;16:536-40.
12. Den Hollander J.E., Erasmus D.A. *Schistosoma mansoni*: male stimulation and DNA synthesis by the female. Parasitology 1985;91 (Pt 3):449-57.
13. Clough E.R. Morphology and reproductive organs and oogenesis in bisexual and unisexual transplants of mature *Schistosoma mansoni* females. J Parasitol 1981;67:535-9.
14. Chen L.L., Rekosh D.M., LoVerde P.T. *Schistosoma mansoni* p48 eggshell protein gene: characterization, developmentally regulated expression and comparison to the p14 eggshell protein gene. Mol Biochem Parasitol 1992;52:39-52.
15. Grevelding C.G., Sommer G., Kunz W. Female-specific gene expression in *Schistosoma mansoni* is regulated by pairing. Parasitology 1997;115 (Pt 6):635-40.
16. Siegel D.A., Tracy J.W. *Schistosoma mansoni*: influence of the female parasite on glutathione biosynthesis in the male. Exp Parasitol 1989;69:116-24.
17. Haseeb M.A., Eveland L.K., Fried B. The uptake, localization and transfer of [4-14C]cholesterol in *Schistosoma mansoni* males and females maintained in vitro. Comp Biochem Physiol A 1985;82:421-3.
18. Basch P.F. Single sex schistosomes and chemical messengers. Parasitology Today 1990;6:160-163.

19. Morand S., Muller-Graf C.D. Muscles or testes? Comparative evidence for sexual competition among dioecious blood parasites (Schistosomatidae) of vertebrates. Parasitology 2000;120 (Pt 1):45-56.

20. Cornford E.M., Huot M.E. Glucose transfer from male to female schistosomes. Science 1981;213:1269-1271.

21. Popiel I., Basch P.F. *Schistosoma mansoni*: cholesterol uptake by paired and unpaired worms. Exp Parasitol 1986;61:343-7.

23. Menrath M., Michel A., Kunz W. A female-specific cDNA sequence of *Schistosoma mansoni* encoding a mucin-like protein that is expressed in the epithelial cells of the reproductive duct. Parasitology 1995;111 (Pt 4):477-83.

24. Schussler P., Potters E., Winnen R., Bottke W., Kunz W. An isoform of ferritin as a component of protein yolk platelets in *Schistosoma mansoni*. Mol Reprod Dev 1995;41:325-30.

25. Schussler P., Kohrer K., Finken-Eigen M., Michel A., Grevelding C.G., Kunz W. A female-specific cDNA sequence of *Schistosoma mansoni* encoding an amidase that is expressed in the gastrodermis. Parasitology 1998;116 (Pt 2):131-7.

26. Tielens A.G., van den Heuvel J.M., van den Bergh S.G. Continuous synthesis of glycogen by individual worm pairs of *Schistosoma mansoni* inside the veins of the final host. Mol Biochem Parasitol 1990;39:195-201.

27. Becker W. [The activation of miracidia of *Schistosoma mansoni* in the eggshell: uptake of water and the metabolic changes connected with it (author's transl)]. Z Parasitenkd 1977;52:69-79.

28. Bobek L.A., Rekosh D.M., LoVerde P.T. Small gene family encoding an eggshell (chorion) protein of the human parasite *Schistosoma mansoni*. Mol Cell Biol 1988;8:3008-16.

29. Hirai H., Tanaka M., LoVerde P.T. *Schistosoma mansoni*: chromosomal localization of female-specific genes and a female-specific DNA element. Exp Parasitol 1993;76:175-81.

30. Reis M.G., Kuhns J., Blanton R., Davis A.H. Localization and pattern of expression of a female specific mRNA in *Schistosoma mansoni*. Mol Biochem Parasitol 1989;32:113-9.

31. Hoffmann K.F., Johnston D.A., Dunne D.W. Identification of *Schistosoma mansoni* gender-associated gene transcripts by cDNA microarray profiling. Genome Biol 2002;3:RESEARCH0041.

32. Eshete F., LoVerde P.T. Characteristics of phenol oxidase of *Schistosoma mansoni* and its functional implications in eggshell synthesis. J Parasitol 1993;79:309-17.

33. El-Sayed N.M.A., Bartholomeu D., Ivens A., Johnston D.A., Lo Verde P.T. Advances in schistosome genomics. Trends Parasitol 2003;in press.

34. Christianson A.M., King D.L., Hatzivassiliou E., Casas J.E., Hallenbeck P.L., Nikodem V.M., Mitsialis S.A., Kafatos F.C. DNA binding and heteromerization of the Drosophila transcription factor chorion factor 1/ultraspiracle. Proc Natl Acad Sci U S A 1992;89:11503-7.

35. Freebern W.J., Osman A., Niles E.G., Christen L., LoVerde P.T. Identification of a cDNA encoding a retinoid X receptor homologue from *Schistosoma mansoni*. Evidence for a role in female-specific gene expression. J Biol Chem 1999;274:4577-85.

36. Mangelsdorf D.J., Evans R.M. The RXR heterodimers and orphan receptors. Cell 1995;83:841-50.

37. Mangelsdorf D.J., Thummel C., Beato M., Herrlich P., Schutz G., Umesono K., Blumberg B., Kastner P., Mark M., Chambon P., et al. The nuclear receptor superfamily: the second decade. Cell 1995;83:835-9.

38. Escriva H., Safi R., Hanni C., Langlois M.C., Saumitou-Laprade P., Stehelin D., Capron A., Pierce R., Laudet V. Ligand binding was acquired during evolution of nuclear receptors. Proc Natl Acad Sci U S A 1997;94:6803-8.

39. Freebern W.J., Niles E.G., LoVerde P.T. RXR-2, a member of the retinoid x receptor family in *Schistosoma mansoni*. Gene 1999;233:33-8.

40. de Mendonca R.L., Escriva H., Bouton D., Zelus D., Vanacker J.M., Bonnelye E., Cornette J., Pierce R.J., Laudet V. Structural and functional divergence of a nuclear receptor of the RXR family from the trematode parasite *Schistosoma mansoni*. Eur J Biochem 2000;267:3208-19.

41. Fantappie M.R., Freebern W.J., Osman A., LaDuca J., Niles E.G., LoVerde P.T. Evaluation of *Schistosoma mansoni* retinoid X receptor (SmRXR1 and SmRXR2) activity and tissue distribution. Mol Biochem Parasitol 2001;115:87-99.

42. de Mendonca R.L., Bouton D., Bertin B., Escriva H., Noel C., Vanacker J.M., Cornette J., Laudet V., Pierce R.J. A functionally conserved member of the FTZ-F1 nuclear receptor family from *Schistosoma mansoni*. Eur J Biochem 2002;269:5700-11.

43. Schussler P., Grevelding C.G., Kunz W. Identification of Ras, MAP kinases, and a GAP protein in *Schistosoma mansoni* by immunoblotting and their putative involvement in male-female interaction. Parasitology 1997;115 (Pt 6):629-34.

44. Osman A., Niles E.G., LoVerde P.T. Characterization of the Ras homologue of *Schistosoma mansoni*. Mol Biochem Parasitol 1999;100:27-41.

45. Kunz W., Gohr L., Grevelding C., Schussler P., Sommer G., Menrath M., Michel A. *Schistosoma mansoni*: control of female fertility by the male. Mem Inst Oswaldo Cruz 1995;90:185-9.

46. Vermeire J.J., Osman A., Lo Verde P.T., Williams D.L. Characterization of the Rho homologue of *Schistosoma mansoni*. Int J Parasitol 2003;33:721-731.

47. Vicogne J., Pin J.P., Lardans V., Capron M., Noel C., Dissous C. An unusual receptor tyrosine kinase of *Schistosoma mansoni* contains a Venus Flytrap module. Mol Biochem Parasitol 2003;126:51-62.

48. Knobloch J., Winnen R., Quack M., Kunz W., Grevelding C.G. A novel Syk-family tyrosine kinase from *Schistosoma mansoni* which is preferentially transcribed in reproductive organs. Gene 2002;294:87-97.

49. Massague J. TGF-beta signal transduction. Annu Rev Biochem 1998;67:753-91.

50. Kloos D.U., Choi C., Wingender E. The TGF-beta--Smad network: introducing bioinformatic tools. Trends Genet 2002;18:96-103.

51. Beall M.J., McGonigle S., Pearce E.J. Functional conservation of *Schistosoma mansoni* Smads in TGF-beta signaling. Mol Biochem Parasitol 2000;111:131-42.

52. Osman A., Niles E.G., LoVerde P.T. Identification and characterization of a Smad2 homologue from *Schistosoma mansoni*, a transforming growth factor-beta signal transducer. J Biol Chem 2001;276:10072-82.

53. Osman A., Niles E.G., Lo Verde P.T. SmSmad4, a co-Smad homologue in *Schistosoma mansoni* responsible for Transforming growth factor-beta (TGF-beta) signaling regulation. J Biol Chem; Submitted.

54. Verjovski-Almeida S., DeMarco R., Martins E.A., Guimaraes P.E., Ojopi E.P., Paquola A.C., Piazza J.P., Nishiyama M.Y., Kitajima J.P., Adamson R.E., Ashton P.D., Bonaldo M.F., Coulson P.S., Dillon G.P., Farias L.P., Gregorio S.P., Ho P.L., Leite R.A., Malaquias L.C., Marques R.C., Miyasato P.A., Nascimento A.L., Ohlweiler F.P., Reis E.M., Ribeiro M.A., Sa R.G., Stukart G.C., Soares M.B., Gargioni C., Kawano T., Rodrigues V., Madeira A.M., Wilson R.A., Menck C.F., Setubal J.C., Leite L.C., Dias-

Neto E. Transcriptome analysis of the acoelomate human parasite *Schistosoma mansoni.* Nat Genet 2003;35:148-57.

55. Davies S.J., Shoemaker C.B., Pearce E.J. A divergent member of the transforming growth factor beta receptor family from *Schistosoma mansoni* is expressed on the parasite surface membrane. J Biol Chem 1998;273:11234-40.

56. Beall M.J., Pearce E.J. Human transforming growth factor-beta activates a receptor serine/threonine kinase from the intravascular parasite *Schistosoma mansoni.* J Biol Chem 2001;276:31613-9.

58. Amiri P., Locksley R.M., Parslow T.G., Sadick M., Rector E., Ritter D., McKerrow J.H. Tumour necrosis factor alpha restores granulomas and induces parasite egg-laying in schistosome-infected SCID mice. Nature 1992;356:604-7.

59. Cheever A.W., Poindexter R.W., Wynn T.A. Egg laying is delayed but worm fecundity is normal in SCID mice infected with *Schistosoma japonicum* and *S. mansoni* with or without recombinant tumor necrosis factor alpha treatment. Infect Immun 1999;67:2201-8.

60. Davies S.J., Grogan J.L., Blank R.B., Lim K.C., Locksley R.M., McKerrow J.H. Modulation of blood fluke development in the liver by hepatic CD4+ lymphocytes. Science 2001;294:1358-61.

61. Wolowczuk I., Nutten S., Roye O., Delacre M., Capron M., Murray R.M., Trottein F., Auriault C. Infection of mice lacking interleukin-7 (IL-7) reveals an unexpected role for IL-7 in the development of the parasite *Schistosoma mansoni.* Infect Immun 1999;67:4183-90.

62. Wolowczuk I., Roye O., Nutten S., Delacre M., Trottein F., Auriault C. Role of interleukin-7 in the relation between *Schistosoma mansoni* and its definitive vertebrate host. Microbes Infect 1999;1:545-51.

63. Wahab M.F., Warren K.S., Levy R.P. Function of the thyroid and the host-parasite relation in murine schistosomiasis mansoni. J Infect Dis 1971;124:161-71.

64. Saule P., Adriaenssens E., Delacre M., Chassande O., Bossu M., Auriault C., Wolowczuk I. Early variations of host thyroxine and interleukin-7 favor *Schistosoma mansoni* development. J Parasitol 2002;88:849-55.

65. Erasmus D.A. "The Adult Schistosome: Structure and Reproductive Biology." In *The Biology of Schistosomes, from Genes to Latrines,* D. Rollinson, A. J. Simpson, eds. London, New York, Sydney, Tokyo, Toronto.: Academic Press, 1987:51-82.

66. Erasmus D.A. *Schistosoma mansoni*: development of the vitelline cell, its role in drug sequestration, and changes induced by Astiban. Exp Parasitol 1975;38:240-56.

Chapter 4

SCHISTOSOMA GENOMICS

David L. Williams[1] and Raymond J. Pierce[2]

[1]*Department of Biological Sciences, Illinois State University, Normal IL 61790-4120;* [2]*Inserm U 547, Institut Pasteur, Lille, France*

Key words: genome sequencing, gene discovery, phylogeny, evolutionary biology, functional genomics, drug and vaccine targets

1.   INTRODUCTION

Schistosomiasis is an important tropical parasitic disease with more than two hundred million human infections and many millions of veterinary infections in more than 70 countries (Chitsulo et al., 2000; Colley et al., 2001). The mortality due to schistosome infections in sub-Saharan Africa is estimated to be 280,000 per year and an additional 20 million individuals suffer severe disease symptoms (van der Werf et al., 2003; WHO Report: www.who.int/tdr). Furthermore, transmission rates have changed little over the past 50 years in spite of the widespread use of chemotherapy and other control strategies. Currently a single drug, praziquantel, is in widespread use with no alternative available should the need arise due to development of drug resistant parasites. Vaccines for schistosomiasis are under development, but none to date promise high levels of efficacy (Bergquist and Colley, 1998; Wilson and Coulson, 1998; Pearce 2003). The development of new chemotherapeutic agents and vaccines will depend on deeper understanding of basic parasite biology and interactions with its host. Genome analysis will provide critical information in the development of disease control strategies as most of these will focus on parasite proteins encoded in the genome. Comparative genomics will also be an important consideration of trematodes genome studies. Platyhelminths are early diverging Bilateria. Phylogeny based on nucleic acid and protein sequences places the trematodes in the Lophotrochozoa (Adoutte et al., 2000). Schistosomes are currently the only member of Lophotrochozoa to be targeting for genome sequencing and analysis. Genomic studies on schistosomes will provide useful information on animal evolution and the development of bilaterian body plan. A better understanding of chromosome structure and evolution will also be produced. Since 1994, the Schistosoma Genome Network (SGN), an international collaboration of laboratories, has received support from the WHO/UNDP/World Bank Special Program for Research and Training in Tropical Diseases (TDR) to study schistosome genomics (Johnston et al., 1999). The network is coordinated by Dr PT LoVerde (State University of New York at Buffalo) with the secretariat of Dr DA Johnston (The Natural History Museum, London UK) and has permanent website (http://www.nhm.ac.uk/hosted_sites/schisto/).

2.   GENOME STRUCTURE

The genome of *Schistosoma* is approximately 270 Mbp (Simpson *et al.*, 1982), which is considerably larger than *Caenorhabditis elegans* (~100 Mbp), *Drosophila melanogastor* (~117 Mbp), and *Ciona intestinalis* (~155 Mbp), similar to *Anopheles gambiae* (~278 Mbp), and about one tenth the size of the human genome. It is estimated that the *S. mansoni* genome has a GC content of 34% (Hillyer, 1974), with 4-8% highly repetitive sequence, 32-36% middle repetitive sequence and 60% single copy sequence (Simpson *et al.*, 1982). Numerous highly or moderately repetitive elements have been identified and their occurrence within existing sequence datasets also indicates that the genome contains at least 30% repetitive sequence (LePaslier *et al.*, 2000). The most important and widespread of the repetitive sequences appear to be retrotransposable elements, described in Chapter 2 of this volume.

Schistosoma are diploid organisms possessing 7 pairs of chromosomes and one pair of sex chromosomes, with ZW sex determination (female=ZW, male=ZZ) (Short and Grossman, 1981). Chromosome size ranges from 18 to 73 Mb and chromosomes can be distinguished by differences in size, arm ratios and C banding pattern (Grossman et al., 1981). C-banding appears as brighter bands stained chromosome spreads and is useful for identifying hybridizing signal locations in fluorescent in situ hybridization mapping (Hirai and LoVerde, 1995).

The number of genes in several animal phyla are known from genome sequences including *Caenorhabditis elegans*, ~19,000 (The *C. elegans* Sequencing Consortium, 1998); *Anopheles gambiae*, 13,683 (Holt et al., 2002); *Drosophila melanogastor*, ~13,472 (Adams et al., 2000); *Ciona intestinalis*, ~15,852 (Dehal et al., 2002). There are estimated to be 15-

20,000 expressed genes in schistosomes (Franco *et al.*, 1995). Considering the simple body plan of schistosomes, is this estimate of gene number reasonable? If one considers that schistosomes are adapted to live in three distinct environments – a warm-blooded vertebrate host, a poikilothermic invertebrate host, and free-living infectious stages – then the estimated number of genes may reflect this adaptation.

3. MITOCHONDRIAL GENOME

The mitochondrial (mt) genomes of schistosomes were an early target for sequencing studies because of their value in species identification, phylogeny and biogeography. Complete or nearly complete mt genome sequences are now available for 11 species or strains of parasitic flatworms (trematodes or cestodes) (Le et al, 2002; McManus et al., 2004). The organization of these genomes is not very different from those of other metazoans apart from the absence of one gene, atp8, which is present in mt genomes from most other phyla. The gene order in most flatworms is similar to that of other protostomes, and in particular to lophotrochozoans such as annelids. However, the usefulness of the platyhelminth mt genome sequences for reconstructing metazoan phylogeny is obscured by their variable base composition and in the case of *S. mansoni* by a drastically altered gene order. Among *Schistosoma* spp., long non-coding regions are rich in repeats and variations in the length of these regions have been observed between individual parasites. The data have been utilized recently as part of a comprehensive study of schistosome phylogeny and biogeography (Lockyer et al., 2003). This study notably confirms the Asian origin of the *Schistosoma*, followed by subsequent dispersal throughout Africa and India, first proposed by Snyder & Loker (2000).

4. GENE DISCOVERY PROJECTS

Because the focus of much work on schistosomiasis is on the identification of new drug and vaccine targets, initial stages of the SGN was on gene discovery by expressed sequence tag (EST) approach (Adams et al., 1991). Partial sequencing of randomly selected cDNA clones generates tags of 100-500 bases for that clone that are screened against gene and protein databases in an attempt to identify the gene. By producing cDNA libraries from multiple stages of the lifecycle, a snapshot of developmental gene expression can be generated. Significant progress in EST generation has been made for both *S. mansoni* and *S. japonicum* in labs from many nations and several papers describing results have appeared (Franco et al., 1995; Franco et al., 1997; Rabelo et al., 1997; Santos et al., 1999; Franco et al., 2000; Fan et al., 1998; Fung et al., 2002). Assembly analysis of all 16,813 *S. mansoni* ESTs has identified 1,763-1,920 tentative consensus sequences (TCs) and 5,570-6000 nonoverlapping ESTs (singletons) depending on the method of clustering utilized (Merrick et al., 2003; Prosdocimi et al., 2002). These sequences are archived in the publicly accessible *S. mansoni* Gene Index (SmGI) at TIGR (http://www.tigr.org/tdb/tgi/smgi/). The ability to utilize the EST databases for gene analysis and discovery is greatly aided by publicly available databases and bioinformatics tools (Oliveira and Johnston, 2001). This can be illustrated by a recent study of a Rho GTPase in *S. mansoni* (Vermeire et al., 2003). The Rho GTPase is a member of a family of small GTP-binding protein that are involved in a number of cell signaling pathways with effects on actin cytoskeleton organization, gene transcription, cell cycle progression, and membrane trafficking. The Rho GTPase family is divided into two subfamilies; the Rho family, which includes RhoA-C and related proteins, and the Rac family, which includes Rac, Cdc42, and related proteins. A query of the SmGI with the Rho GTPase identified by Vermeire and coworkers revealed the presence of two additional Rho genes, as well as orthologues of Cdc42 and Rac. With the sequences of these genes in hand, their expression and encoded proteins can now be more easily studied.

A large-scale multi-center gene discovery project has been initiated in the state of Sao Paolo, Brazil to sequence EST clones generated by the ORESTES low-stringency RT-PCR amplification technique (Dias-Neto et al., 1997; Dias-Neto et al., 2000). The ORESTES approach is based on arbitrary primers and low stringency RT-PCR and preferentially amplifies the central portion of messages which contain the more conserved, function-defining coding regions (Fietto et al., 2002). This project started in April, 2001 and has reached its goal of generating over 160,000 ESTs from six selected stages of the life cycle, cercariae, seven-day cultured schistosomula, adult worms, eggs, miracidia and germ balls (Verjovski-Almeida et al., 2003). The project has acquired over 68 million high-quality bases and all stages were covered approximately equally. Using cluster analysis it was determined that there were 18 million non-redundant bases in accepted sequences, resulting in an average redundancy (acquired bases/non-redundant bases) of 3.8 X. A web server based on comparison to multiple public databases has been developed and on completion of the analysis the server tools and sequence database will be publicly available (http://verjo18.iq.usp.br/schisto/). The data should prove to be extremely useful in gene discovery and genome annotation.

The SGN has mainly focused on *S. mansoni*. However, there is a growing realization that various schistosome species differ in many biological aspects, including morphological characters, infectivity to snails, range of definitive host, growth rates, egg production, prepatency periods, pathogenicity, and immunogenicity (Le et. al., 2000). This has prompted the development of a number of projects to explore the genomes of other schistosome species as well as related taxa. The Chinese National Human Genome Center at Shanghai (CHGCS) in cooperation with Shanghai Institute of Parasitology, Chinese Academy of Medical Science and Shanghai Second Medical University, have completed a large-scale *S. japonicum* EST project (Hu et al., 2003). To date, 43,707 available ESTs have been derived from male and female adult worms and eggs (total of 45,900 entries in dbEST (release 053003) for *S. japonicum*). These ESTs have been integrated into 13,131 clusters, of which 8,523 (64.91 %) displayed homology with genes identified or predicted in other organisms in the public database. Moreover, 610 full-length cDNAs were also obtained. Interestingly, *S. japonicum*, in common with *S. mansoni*, seems to share a number of growth factor receptors and nuclear hormone receptors with its mammalian hosts, supporting the hypothesis that schistosomes could utilize signals from its hosts for development and maturation (see de Mendonça et al, 2000a for review).

With Wellcome Trust funding, collaboration is underway at The Sanger Institute to generate 15,000 ESTs of *S. haematobium* and *Fasciola hepatica* from cDNA libraries representing diverse stages of the life cycles (http://www.sanger.ac.uk/Projects/S_mansoni/). *Fasciola hepatica* has been selected as the platyhelminth of choice for initial comparative EST analysis because it is a digenean and is phylogenetically reasonably close to *Schistosoma*. Like schistosomes *F. hepatica* has a two-host life cycle and it is an economically important human and veterinary pathogen. Data from *F. hepatica* will form a framework for future studies on *F. gigantica*, another parasite of direct practical and economic relevance to the developing world.

5.   PHYSICAL MAPPING

As a target for genome analysis, *Schistosoma* provides particular challenges. Schistosomes possess the largest genome of any of the parasites currently targeted for genome sequencing and analysis (at 270 Mbp, its genome is some 3 times that of *Brugia malayi*, and 10 times that of *P. falciparum*). The genome cannot be partitioned by pulsed field gel electrophoresis or flow cytometry to permit a chromosome-by-chromosome analysis. Therefore, for genome sequencing to become a reality a physical map, both to identify the minimal tile path and to provide a scaffold to guide whole genome sequence assembly, will be required. Genetic mapping is not a realistic prospect as technical difficulties preclude widespread application of crossing experiments (Pica-Mattoccia et al. 1993).

As envisaged by the SGN a physical map of the *S. mansoni* genome would be composed of contigs of overlapping large-insert genomic clones spanning the entire genome. The first stages in the development of a physical map saw the generation of yeast artificial chromosome (YAC) and bacterial artificial chromosome (BAC) libraries. An *S. mansoni* YAC library has been constructed containing more than 2,283 clones with and average insert size of 358 kbp providing 2.6-fold genome coverage (Tanaka et al., 1995). One hundred randomly selected YAC clones were assigned to the karyotype by fluorescent *in situ* hybridization (FISH) and a first generation chromosome map has been produced (Hirai and LoVerde, 1995). However, this preliminary map indicated that the YAC library displayed a notable cloning bias toward the rRNA gene repeat region of chromosome 2 and numerous clones were localized to more than one chromosomal site suggesting the presence of chimeric clones (Tanaka et al., 1995). In order to complement the *S. mansoni* YAC library, to provide more uniform genome coverage, and to facilitate DNA preparation a BAC library was prepared from cercarial DNA (Le Paslier et al., 2000). This library (Sm1) consists of more than 21,000 recombinant clones in the pBeloBAC11 vector. The mean insert size was 100 kbp representing about 8-fold coverage of the genome. The Sm1 BAC library has better genome coverage than the YAC library. Another BAC library has been constructed by Dr. Philip T. LoVerde and The Institute for Genomic Research (TIGR) in collaboration with Dr. Pieter de Jong at the Children's Hospital Oakland Research Institute. This library (CHORI-103) used a different restriction enzyme to generate DNA fragments (*Bam* HI instead of *Hind* III) and the vector pTARBAC1.3 to increase representative genome coverage. This library consists of 34,708 clones with a mean insert size of 140 kbp representing 19-fold coverage of the genome.

With support from WHO/TDR, physical mapping of chromosome 3 by BAC hybridization was undertaken. Using YAC and BAC clones mapped to chromosome 3 by FISH as entry points, several BAC contigs have been assembled. Because of the repetitive nature of the *Schistosoma* genome, hybridization-based approaches with entire large insert clones for contig assembly could not be used. Instead, BAC end sequences were generated, screened for repetitive elements, PCR primers were designed to unique regions of the end sequence and used to generate probes for the next round of screening. This approach has proven to be cumbersome and the abundance of repetitive DNA has impeded screening making this approach impractical for full genome mapping (Williams, Pierce, unpublished). Instead, alternative mapping strategies will be pursued with funds recently obtained from the Wellcome Trust. The goals of this project are to generate, and make available to the public, an integrated HAPPY map and BAC tile path for the genome *S. mansoni* as a resource to (i) underpin future genome sequencing efforts by allowing rapid and efficient sequencing and assembly and (ii) provide a rapid means of isolating genomic copies of genes of interest (Foster and Johnston, 2002). Because of its large and complex genome a physical map/tile path of the *S. mansoni* genome represents an essential scaffold to permit rapid and efficient sequencing and assembly at the whole genome, single chromosome or targeted region level.

HAPPY mapping is a physical analogue of meiosis (Fig. 4-1), comparable to classical genetic mapping (pedigree analysis), but measuring the frequency of induced breaks between markers, rather than recombination frequency between markers after meiosis (Dear and Cook, 1993; Dear *et al.*, 1998).

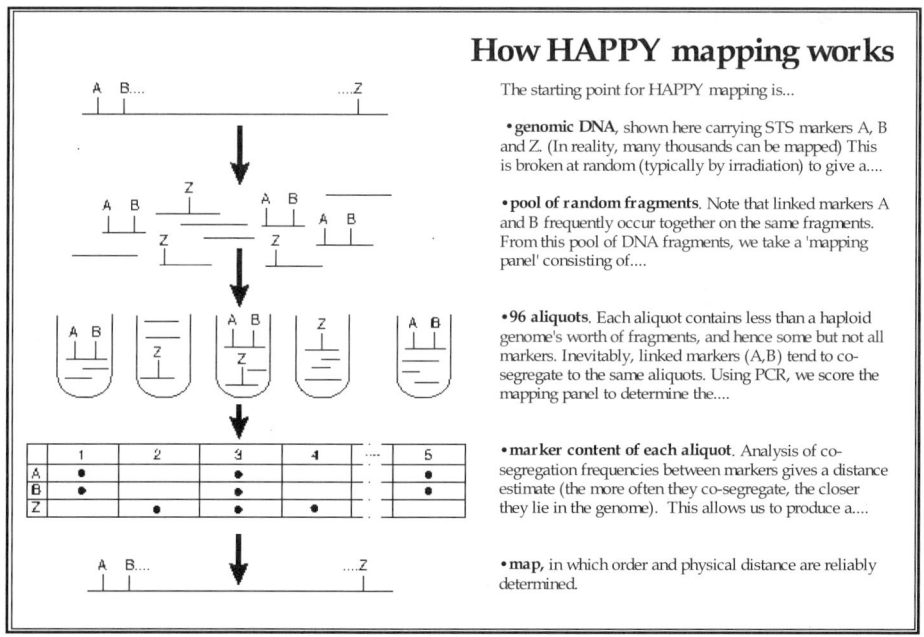

Figure 4-1. Description of HAPPY mapping methodology.

6.   GENOME SEQUENCING

A first step in genome sequencing strategy has been to sequence 500 bp each from the ends of randomly selected clones from two BAC libraries. The end sequences will enhance early gene discovery, serve as markers for the construction of the high-resolution sequence-ready maps, and provide long-range scaffolds to assemble whole genome shot gun sequences. BAC end sequencing has been done at Genoscope in Paris and at TIGR. Some 14,750 BAC end sequences from the Sm1 BAC library have been generated by Genoscope and 12,576 and 19,199 BAC end sequences from the Sm1 and CHORI-103 libraries respectively have been produced by TIGR. These efforts have produced ~28 Mbp (about 0.1 X genome coverage) of discontinuous single-pass sequence. The sequences generated from the Sm1 BAC library have been deposited in dbGSS at NCBI while the CHORI-103 end sequences are available through the TIGR web/ftp sites.

7. BAC SEQUENCING

With support form the NIH and the Wellcome Trust respectively, TIGR and The Sanger Institute have shotgun sequenced a number of BAC clones. This will help to determine the effect of repetitive elements on sequence and map assembly, to understand higher level genome organization including: the distribution, arrangement and flanking regions of repetitive elements, gene structure and organization, upstream and downstream gene regulatory elements, the length and composition of 5' and 3' UTRs, and to provide calibrated reference data for HAPPY mapping. Sequencing large insert clones such as BACs involves first making sublibraries from randomly sheared and size-selected (~2 kb) fragments in small insert libraries. The ends of ~1000 clones from the sublibraries per BAC are sequenced and the end sequences are assembled to span the BAC insert. Thus far nine (769,994 bp) and ten (~1.1 Mbp) BACs have been sequenced at TIGR (http://www.tigr.org/tdb/e2k1/sma1/) and The Sanger Institute (http://www.sanger.ac.uk/Projects/S_mansoni/) respectively. Because TIGR selected BACs that had been mapped by FISH to chromosomal location, these BACs would also serve as nucleation sites for a map-as-you-go strategy for sequencing along the length of the chromosome. The sequenced BAC clone would be used to identify new overlapping BAC clones by virtue of their end sequences identified from the BAC end sequence database. The restriction fragment fingerprints of a selection of the overlapping BAC clones would be compared to identify any BACs containing artifacts or inconsistencies so that they will be eliminated as sequence substrates and to identify the minimum sequencing tiling path (minimum overlap of BACs). Selected BACs would then be sequenced extending outwards from seed clones to develop contigs of BAC clones (Venter et al., 1996). An advantage to this approach is that is does not require the construction of physical maps.

8.   WHOLE GENOME SHOTGUN SEQUENCING

At least in theory, starting from multiple entry points this map-as-you-go strategy could be used to sequence the entire genome. However, it was soon discovered to be difficult if not impossible to map out from sequenced BACs using the BAC end sequence database. A potential reason for this is that BAC end sequences may not be randomly distributed along the genome but concentrated in the repeat sequences (El Sayed, Williams personal comm.). This has prompted a change in strategy to whole genome shotgun (WGS) sequencing (Fig. 4-2). The WGS sequencing approach is helped by the recent success of WGS sequencing in other organisms and substantial reductions in the overall costs of sequencing. Because of the difficulty in separating schistosome chromosomes, libraries of sheared (~2 kbp) whole genomic DNA in pUC18-based, low copy plasmids will provide most of the sequencing templates. These clones will be sequenced from both ends to produce pairs of linked sequences representing ~500 bp at the ends of each insert. End sequences from a second library containing ~10 kbp will provide medium-range linking, including spanning the abundant repeat elements, while BAC end sequences will provide higher order linkage and translation to the HAPPY map. Starting in late 2002 with NIH support and in mid 2003 supported by the Wellcome Trust, TIGR and The Sanger Institute plan generate approximately 8X sequence coverage and identification of 97% of the genes by the fall of 2003. Problems predicted with schistosome genome assembly and annotation include: large % of highly conserved repeats in the genome, libraries made from outbred organisms that display significant genetic diversity, and the small number of sequenced genes available to train gene-finding software. Successful completion of the *Ciona intestinalis* genome, an organism with similar limitations, is encouraging. As of mid-September 2003, 1,101,469 sequence reads (~800,000,000 bp or ~3 x coverage) and 583,300 sequence reads (413,726,285, 1.5 x coverage) have been generated by TIGR and The Sanger Institute, respectively. At this level of coverage, it is estimated that less than 2% of the *S. mansoni* genome remains unsequenced.

Recently, the CHGCS has been developing a draft genomic sequence of *S. japonicum*, in an attempt to accelerate gene discovery and uncover the basic features of *S. japonicum* genome. Genome sequencing approaches are similar to those being utilized for *S. mansoni*. The first BAC library of *S. japonicum* has been constructed by the CHGCS in collaboration with Dr. Pieter de Jong employing the Anhui strain. The average insertion size of the BAC clones is around 158 kb representing about 20 X coverage of the *S. japonicum* genome. A high-resolution physical map will be constructed based on BAC end sequencing, fingerprinting and FISH analysis.

Meanwhile, a draft sequence of 5-6 fold coverage of the *S. japonicum* haploid genome will be obtained by using the shotgun sequencing method. The annotation and analysis of genomic sequence, as well as on the ESTs of *S. japonicum* will help to explore the characteristics of its genome, to identify expressed genes and their biological functions. The endeavor will make a substantial contribution to the understanding of the evolution of helminths and the elucidation of host-parasite interaction.

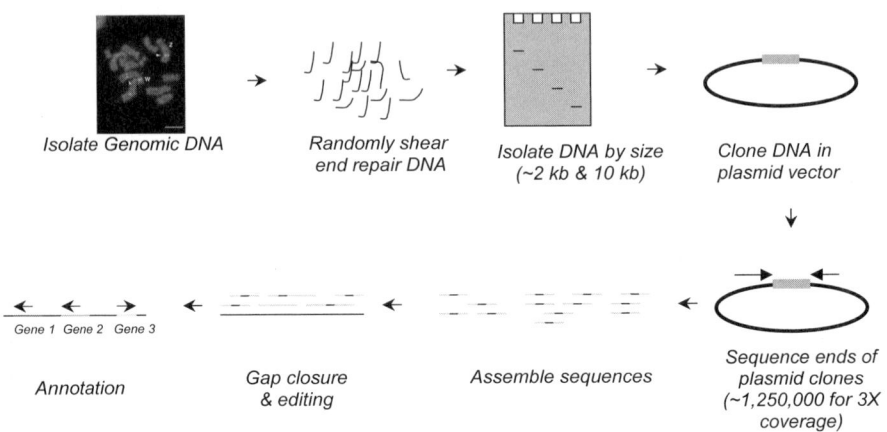

Figure 4-2. Schema for genome shotgun sequencing. (FISH image courtesy H. Hirai).

9.   FUNCTIONAL GENOMICS – USES OF THE GENOME SEQUENCE INFORMATION

The genome sequence will be an invaluable asset in the study of schistosomes. In addition to the identification of most coding regions, gene order/higher order chromosome structure, and various repeats will be described. The functional analysis of the encoded genes will remain as a major task. The first approach for the functional assignment of coding region will be by homology to known genes from other organisms using Blast and FASTA programs. Proteins will be sorted into functional classes based on the Munich Information Centre for Protein Sequences (MIPS) catalogue (Mewes et al., 2002) and screened for predicted transmembrane segments, GPI addition peptide signal sequence (membrane proteins) and N-terminal signal/targeting sequences (secreted or glandular proteins) using a variety of bioinformatics tools.

It is expected that a significant number of genes will encode proteins of unknown function. For instance, 60% of predicted peptides in the *Plasmodium falciparum* appear to be unique to that organism (and have unknown function) (Gardner et al., 2002) and even in the well-studied *Escherichia coli*, nearly 40% of the predicted open reading frames encoded proteins of unknown function (Blattner et al., 1997). Therefore, functional assignment by alternative means will be necessary (Table 4-1). One approach to functional assignment, as well as providing fundamental knowledge on the biology of schistosomes, is to study the expression pattern of genes during the parasite lifecycle. Several methods for gene expression analysis exist including EST (described above), cDNA microarrays (Hoffmann et al., 2002, Chapter 7), and serial analysis of gene expression (SAGE).

In SAGE, a short sequence tag (14-16 bp or 20-21 bp) from a unique position of a mRNA molecule is used to uniquely identify the source gene from within the genome (Velculescu et al., 1995; Saha et al., 2002). The population of tags defines patterns of expression of individual genes. Quantification of all tags provides a relative measure of gene expression (i.e. mRNA abundance). SAGE thus provides both the identity of expressed genes and levels of their expression. Such a combination of SAGE expression data and a known genome sequence provides the ability to confirm ORF predictions, detect unpredicted ORFs, and annotate a complete genome with gene expression information (Velculescu et al., 2000; Saha et al., 2002). SAGE has been used to monitor gene expression during the yeast cell cycle (Velculescu et al. 1997) and for investigations of human disease and for surveys of gene expression and cellular differentiation in a variety of human and mammalian cells and tissues (Velculescu et al., 1999; http://www.sagenet.org/). SAGE analysis of the transcriptome of the malarial parasite *Plasmodium falciparum* revealed that 70% of highly abundant SAGE tags matched unique loci, detected uncharacterized open reading frames, demonstrated significant production of antisense transcripts, and identified high expression of genes encoding important membrane associated proteins (Munasinghe et al. 2001; Patankar et al. 2001).

Although primary genomic sequence data will provide many insights to parasite biology, it will be of limited value without characterizing the expression, tissue and sub/cellular localization or secretion of proteins throughout the parasite's complex life cycle. Proteomics, the analysis of the parasite's total protein complement (the proteome), will be essential to understand the function of many schistosome genes identified from the genome sequence, will provide invaluable information to aid in the annotation of *S. mansoni* genome sequence, and will identify drug and vaccine targets, which are likely to be proteins. Functional proteomic analysis will also integrate information on gene transcription and mRNA translation two events not necessarily coupled in *S. mansoni* (Pierrot et al., 1996). Recent advances in proteome analysis have been possible with the application of two-dimensional gel electrophoresis and mass spectrometry (MS) (reviewed in Yates, 2000; Ashton et al., 2001; Hunter et al., 2002, Lin et al., 2003).

In high-throughput or shotgun proteomics, protein mixtures are isolated from selected organelles, tissues, or whole organisms, digested to peptides and analyzed by inline, high resolution liquid chromatography (LC) and tandem mass spectrometry. One example of this analysis is multidimensional protein identification technology (MudPIT) (Link et al., 1999). In this process trypsin-treated protein mixtures are loaded into a biphasic microcapillary column packed in series with a strong cation exchange (SCX) resin followed by a reverse-phase (RP) resin. A discrete fraction of absorbed peptides are eluted from the SCX column and are displaced onto the RP column. Peptides are retained on the RP column and then eluted from the RP column into the mass spectrometer. An iterative process of increasing salt concentration is then used to displace additional fractions of peptides from the SCX column onto the RP column. Each discrete fraction is eluted from the RP column into the mass spectrometer. Because SCX separates by charge and RP separates by hydrophobicity, the two dimensions of chromatography are orthogonal. These methods have allowed the identification of 1,484 proteins from *S. cerevisiae* (Washburn et al., 2001), 2,415 proteins from four stages of the human malaria parasite *P. falciparum* (Florens et al., 2002), and the most comprehensive proteome analysis to date: the identification of 2,528 proteins from rice (*Oryza sativa*) using a combined MudPIT/two-dimensional gel electrophoresis approach (Koller et al., 2002). With MudPIT it should be possible to efficiently analyze schistosome proteomes and protein complements such as worm surface proteins and egg or larval secretions as well as proteomes of various life stages could be individually identified. Further analysis of protein function will be accomplished by the use of gene disruption and transgenesis. Gene disruption by RNA interference (RNAi) uses double-stranded RNA to knock down levels of targeted RNA to generate phenotypes similar to gene-knock outs and test gene function. RNAi studies are beginning to provide exciting information on the study of schistosomes' protein function in both mammalian and molluscan hosts (Boyle et al., 2003; Skelly et al., 2003). Transgenic schistosomes can be generated using biolistic methods (Davis et al., 1999, Wippersteg et al, 2002a; 2002b). These methods can be used to study gene promoter function and structure, the effects of ectopic and overexpression of proteins, and other molecular applications. The potential use of retroelements in transgenesis is described in Chapter 2.

Table 4-1. Summary of strategies for gene identification and functional analysis.

Methodology	Comments
Expressed sequence tags (ESTs)	High-throughput gene identification. Produces tags for gene identification, stage expression, and gene family analysis (Franco et al., 2000).
Serial analysis of gene expression (SAGE)	High-throughput method for measuring genome-wide expression by sampling small sequence tags (Velculescu et al., 1995). Requires small amounts of RNA and no prior sequence knowledge.
cDNA microarray analysis	High-throughput method for measuring genome-wide expression based on hybridization to cDNA microarrays (Hoffmann et al., 2002, Hoffmann and Fitzpatrick, Chapter 7). Can be mass-produced for community use.
Gene disruption	RNAi. Uses double-stranded RNA to knock down levels of targeted RNA to generate phenotypes similar to gene-knock outs. Methods in development for schistosomes (Boyle et al., 2003; Skelly et al., 2003).
Transgenic organisms	Gene-gun based methods have had limited success for the introduction of transgenes into schistosomes (Davis et al., 1999, Wippersteg et al, 2002a; 2002b). Potential use of retroelements in development (Brindley, et al., Chapter 2).
Proteome analysis	Two-dimensional gel electrophoresis and mass spectrometry. Identification of proteins separated by 2D PAGE and analysis by ms and comparison of peptide mass fragments to theoretical pmf predicted from genome database (Ashton et al., 2001).
Multidimensional protein identification technology (MudPIT)	High-throughput proteomic analysis of protein complexes. Tissues, membrane proteins, whole organisms can be analyzed. Analysis by inline, high resolution liquid chromatography (LC) and tandem mass spectrometry (Link et al., 1999).

10.   GENOME SEQUENCING AND EVOLUTIONARY BIOLOGY

10.1   Hox genes and platyhelminth phylogeny

Among the important issues that can be resolved by the acquisition of large-scale genome sequence data is the evolutionary position of the platyhelminths within the phylogenetic tree of the metazoa. Molecular phylogeny, and particularly the use of 18S rRNA sequences has recently led to the division of the protostomes into two main groups or clades, the ecdysozoans and the lophotrochozoans (Aguinaldo et al., 1997; Adoutte et al, 2000). Furthermore, the nematodes and platyhelminths, traditionally considered as early offshoots at the base of the metazoan tree, before the main division between the protostomes and the deuterostomes, were placed within the ecdysozoan and lophotrochozoan clades respectively. This new classification is still highly controversial (Blair et al., 2002) and can only be definitively resolved by the analysis of other genes or gene families.

Hox genes are present in all metazoans except sponges and in triploblastic metazoans, also known as bilateria, are generally grouped together in the Hox cluster. Hox genes encode related homeodomain transcription factors that are involved in determining organization of the anterior-posterior axis, as in the segmental morphology of insects. They also generally respect the colinearity rule; they are expressed along the body axis in the same order as they are found in the cluster. All these characteristics arose early in the history of the metazoa (Finnerty & Martindale, 1998) and therefore the analysis of Hox clusters and the constituent genes has become a key element in the understanding of metazoan phylogeny (de Rosa et al., 1999, Balavoine et al., 2002).

Although individual Hox genes have been characterized from various platyhelminths and other lophotrochozoans, including one potential member of the cluster from *S. mansoni*, Smox 1 (Webster & Mansour, 1992), the analysis of a complete lophotrochozoan cluster and the characterization of its constituent genes has not been done. In a preliminary PCR-based study aimed at characterizing the *S. mansoni* Hox cluster, three further Hox genes have been characterized (Pierce et al., submitted). These genes are located on distinct BAC clones and chromosome walking has failed to link the genes together in a contig. However, the chromosomal location of the BACs has shown that at least three of the four *S. mansoni* Hox genes co-localize to the long arm of chromosome 4, suggesting that an extended cluster may exist in schistosomes. In this respect the platyhelminth cluster may be more similar to those found in *Drosophila* or *C. elegans*, than to those of mammals. Analysis of the genes themselves, however, supports the phylogenetic position of platyhelminths within the Lophotrochozoa. The genome sequence of *S. mansoni*, the only lophotrochozoan currently being sequenced, will help elucidate the Hox cluster organization in this clade.

10.2   The repertoire of nuclear and membrane receptors

Two further gene families that are under close scrutiny both from the point of view of comparative phylogeny, and because they are potential target molecules for novel chemotherapeutic strategies are nuclear receptors (NR) and receptor tyrosine kinases (RTK). Nuclear receptors form a superfamily of ligand-dependent transcription factors that interact with steroid (estrogen, progesterone, androgens, ecdysone) or non-steroid (retinoic acids, vitamin D, prostaglandins) ligands to control gene expression (Gronemeyer & Laudet, 1995). Again using PCR-based strategies, so far six NRs (Escriva et al., 1997; Freebern et al., 1999) have been identified. All of these are in fact members of NR families that are so-called orphan receptors in invertebrates, in that no ligands have been found up till now. Further characterization of three of these receptors, SmRXR1 (Freebern et al., 1999) SmRXR2 (de Mendonça et al., 2000b) and SmFTZ-F1 (de Mendonça et al., 2002) show that SmRXR1 and SmFtz-F1 are show structural and some functional conservation compared to members of the same NR families in vertebrates and arthropods, whereas SmRXR2 is structurally and functionally divergent. Tantalizingly, the *S. japonicum* ESTs contain NRs that seem to belong to families that are ligand-dependent in vertebrates, but these need to be fully characterized before this can be affirmed. For more discussion of schistosome nuclear receptor biology, see Chapter 3.

RTKs are plasma membrane receptors, generally of peptide growth factors. Similar cloning strategies have yielded three *S. mansoni* RTKs, SER (Ramachandran et al., 1996), a homologue of the mammalian EGF receptor, SmRTK2, a homologue of the insulin receptor and SmRTK1 (Vicogne et al, 2003), which has an insulin receptor family tyrosine kinase domain and a highly original extracellular domain homologous to those of the glutamate and GABA receptors. Another membrane growth factor receptor, called SmRK1, is a homologue of the mammalian TGFβ receptor and is a receptor serine/threonine kinase (Beall & Pearce, 2001). SmRK1 is activated by human TGFβ, which lends support to the hypothesis that host hormones and growth factors directly influence schistosome growth and survival.

The presence of both conserved and original NRs and RTKs in *S. mansoni* mirrors the repertoire of these receptor classes in *C. elegans*, in which mining of the genomic sequence data has shown that this nematode has 40 RTKs, of which 29 correspond to known classes and 11 of which appear to be specific to this organism (Plowman et al. 1999). This organism specificity is even more exacerbated when the NRs are examined since *C. elegans* counts no less than 270 genes, compared to only 21 for *Drosophila* (Sluder & Maina, 2001). Although schistosomes may not possess such an exaggerated diversity of receptors, it is already clear that interesting and original molecules will be revealed by both EST and genomic sequencing and that many of these will be potential drug targets. Moreover, the nature of the schistosome nuclear receptors will help to resolve the debate over the evolution of this superfamily; i.e. whether or not the ancestral nuclear receptor was an orphan, or had a ligand (Laudet, 1997).

ACKNOWLEDGEMENTS

We thank David Johnston (NHM, London), Bernard Konfortov and Paul Dear (MRC, Cambridge), Sergio Verjovski-Almeida (University of Sao

Paulo), Najib El Sayed (TIGR), Shengyue Wang, Zheguan Han, Zhen Feng, Gouping Zhao, and Zhu Chen (Chinese National Human Genome Center at Shanghai), and Philip LoVerde (SUNY Buffalo) for helpful discussions and communication of unpublished information.

BIBLIOGRAPHY

Adoutte, A., Balavoine, G., Lartillot, N., Lespinet, O., Prud'homme, B.,and de Rosa, R., 2000, The new animal phylogeny: reliability and implications, *Proc. Natl. Acad. Sci. U S A.* **97**:4453-4456.

Adams, M.D., Kelley, J.M., Gocayne, J.D., Dubnick, M., Polymeropoulos, M.H., Xiao, H., Merril, C.R., Wu, A., Olde, B., Moreno, R.F., et al, 1991, Complementary DNA sequencing: expressed sequence tags and human genome project, *Science* **252**:1651-1656.

Adams, M.D., Celniker, S.E., Holt, et al., 2000, The genome sequence of *Drosophila melanogaster, Science* **287**:2185-2195.

Aguinaldo, A.M., Turbeville, J.M., Linford, L.S., Rivera, M.C., Garey, J.R., Raff, R.A., and Lake, J.A., 1997, Evidence for a clade of nematodes, arthropods and other moulting animals, *Nature* **387**: 489-493.

Ashton, P.D., Curwen, R.S., and Wilson, R.A., 2001, Linking proteome and genome: how to identify parasite proteins, *Trends Parasitol.* **17**:198-202.

Balavoine, G., de Rosa, R., and Adoutte, A., 2002, Hox clusters and bilaterian phylogeny, *Mol. Phylogenet. Evol.* **24**:366-373.

Beall, M.J., and Pearce, E.J., 2001, Human transforming growth factor-beta activates a receptor serine/threonine kinase from the intravascular parasite *Schistosoma mansoni, J. Biol. Chem.* **276**:31613-31619.

Bergquist, N.R., and Colley, D.G., 1998, Schistosomiasis vaccines: research to development, *Parasitol. Today.* **14**:99-104.

Blair, J.E., Ikeo, K., Gojobori, T., and Hedges, S.B., 2002, The evolutionary position of nematodes, *BMC Evol. Biol.* **2**: 1-7. http://www.biomedcentral.com/1471-2148/2/7.

Blattner, F.R., Plunkett, G., 3rd, Bloch, C.A., Perna, N.T., Burland, V., Riley, M., Collado-Vides, J., Glasner, J.D., Rode, C.K., Mayhew, G.F., Gregor, J., Davis, N.W., Kirkpatrick, H.A., Goeden, M.A., Rose, D.J., Mau, B., and Shao, Y., 1997, The complete genome sequence of *Escherichia coli* K-12, *Science* **277**:1453-1474.

Boyle, J.P., Wu, X.J., Shoemaker, C.B., and Yoshino, T.P., 2003, Using RNA interference to manipulate endogenous gene expression in *Schistosoma mansoni* sporocysts, *Mol. Biochem. Parasitol.* **128**:205-215.

Chitsulo, L., Engels, D., Montresor, A., Savioli, L., 2000, The global status of schistosomiasis and its control, *Acta Trop.* **77**:41-51.

Colley, D.G., LoVerde, P.T., and Savioli, L., 2001, Infectious disease. Medical helminthology in the 21st century, *Science* **293**:1437-1438.

Davis, R.E., Parra, A., LoVerde, P.T., Ribeiro, E., Glorioso, G., and Hodgson, S., 1999, Transient expression of DNA and RNA in parasitic helminths by using particle bombardment, *Proc. Natl. Acad. Sci. U S A.* **96**:8687-8692.

De Mendonça, R.L., Escriva, H., Bouton, D., Laudet, V., and Pierce, R.J., 2000a, Hormones and nuclear receptors in schistosome development, *Parasitol. Today* **16**:233-240.

De Mendonça, R.L., Escriva, H., Bouton, D., Zelus, D., Vanacker, J.M., Bonnelye, E., Cornette, J., Pierce, R.J., and Laudet, V., 2000b, Structural and functional divergence of a nuclear receptor of the RXR family from the trematode parasite *Schistosoma mansoni*, *Eur. J. Biochem.* **267**:3208-3219.

De Mendonça, R.L., Bouton, D., Bertin, B., Escriva, H., Noël, C., Vanacker, J.M., Cornette, J., Laudet, V., and Pierce, R.J., 2002, A functionally conserved member of the FTZ-F1 nuclear receptor family from *Schistosoma mansoni*, *Eur. J. Biochem.* **269**:5700-5711.

De Rosa, R., Grenier, J.K., Andreeva, T., Cook, C.E., Adoutte, A., Akami, M., Carroll, S.B., and Balavoine, G., 1999, Hox genes in brachiopods and priapulids and protostome evolution, *Nature* **399**: 772-776.

Dear, P.H., and Cook, P.R., 1993, Happy mapping: linkage mapping using a physical analogue of meiosis, *Nucleic Acids Res.* **21**:13-20.

Dear, P.H., Bankier, A.T., and Piper, M.B., 1998, A high-resolution metric HAPPY map of human chromosome 14, *Genomics* **48**:232-241.

Dehal, P., Satou, Y., Campbell, et al., 2002, The draft genome of *Ciona intestinalis*: insights into chordate and vertebrate origins, *Science* **298**:2157-2167.

Dias Neto, E., Harrop, R., Correa-Oliveira, R., Wilson, R.A., Pena, S.D., and Simpson, A.J., 1997, Minilibraries constructed from cDNA generated by arbitrarily primed RT-PCR: an alternative to normalized libraries for the generation of ESTs from nanogram quantities of mRNA, *Gene* **186**:135-142.

Dias Neto, E., Correa, R.G., Verjovski-Almeida, S., et al., 2000, Shotgun sequencing of the human transcriptome with ORF expressed sequence tags, *Proc. Natl. Acad. Sci. USA* **97**:3491-3496.

Escriva, H., Safi, R., Langlois, M.C., Saumitou-Laprade, P., Stehelin, D., Capron, A., Pierce, R.J., and Laudet, V., 1997, Ligand binding was acquired during evolution of nuclear receptors, *Proc. Natl. Acad. Sci. USA* **94**:6803-6808.

Fan, J., Minchella, D.J., Day, S.R., McManus, D.P., Tiu, W.U., and Brindley, P.J., 1998, Generation, identification, and evaluation of expressed sequence tags from different developmental stages of the Asian blood fluke *Schistosoma japonicum*, *Biochem. Biophys. Res. Commun.* **252**:348-356.

Fietto, J.L.R., DeMarco, R., and Verjovski-Almeida, S., 2002, Use of degenerate primers and touchdown PCR for construction of cDNA libraries, *Biotech.* **32**:1404-1411.

Finnerty, J.R., and Martindale, M.Q., 1998, The evolution of the Hox cluster: insights from outgroups, *Curr. Op. Genet. Dev.* **8**: 681-687.

Florens, L., Washburn, M.P., Raine, J.D., Anthony, R.M., Grainger, M., Haynes, J.D., Moch, J.K., Muster, N., Sacci, J.B., Tabb, D.L., Witney, A.A., Wolters, D., Wu, Y., Gardner, M.J., Holder, A.A., Sinden, R.E., Yates, J.R., and Carucci, D.J., 2002, A proteomic view of the *Plasmodium falciparum* life cycle, *Nature* **419**:520-526.

Foster, J.M., and Johnston, D.A., 2002, Helminth genomics: from gene discovery to genome sequencing, *Trends Parasitol.* **18**:241-242.

Franco, G.R., Adams, M.D., Soares, M.B., Simpson, A.J., Venter, J.C., and Pena, S.D., 1995, Identification of new *Schistosoma mansoni* genes by the EST strategy using a directional cDNA library, *Gene* **152**:141-147.

Franco, G.R., Rabelo, E.M., Azevedo, V., Pena, H.B., Ortega, J.M., Santos, T.M., Meira, W.S., Rodrigues, N.A., Dias, C.M., Harrop, R., Wilson, A., Saber, M., Abdel-Hamid, H., Faria, M.S., Margutti, M.E., Parra, J.C., and Pena, S.D., 1997, Evaluation of cDNA libraries from different developmental stages of *Schistosoma mansoni* for production of expressed sequence tags (ESTs), *DNA Res.* **4**:231-240.

Franco, G.R., Valadao, A.F., Azevedo, V., and Rabelo, E.M., 2000. The *Schistosoma* gene discovery program: state of the art, *Int. J. Parasitol.* **30**:453-463.

Freebern, W.J., Osman, A., Niles, E.G., Christen, L., and LoVerde, P.T., 1999, Identification of a cDNA encoding a retinoid X receptor homologue from *Schistosoma mansoni, J. Biol. Chem.* **274**:4577-4585.

Fung, M.C., Lau, M.T., and Chen, X.G., 2002, Expressed sequence tag (EST) analysis of a *Schistosoma japonicum* cercariae cDNA library, *Acta Trop.* **82**:215-224.

Gardner, M.J., Hall, N., Fung, E., et al., 2002, Genome sequence of the human malaria parasite *Plasmodium falciparum, Nature* **419**:498-511.

Gronemeyer, H., and Laudet, V., 1995, Transcription factors 3: nuclear receptors, in: *Protein Profile*, Vol 2, Academic Press, London, pp. 1173-1308.

Grossman, A.I., Short, R.B., and Cain, G.D., 1981, Karyotype evolution and sex chromosome differentiation in Schistosomes (Trematoda, Schistosomatidae), *Chromosoma* **84**:413-430.

Hillyer, G.V., 1974, Buoyant density and thermal denaturation profiles of schistosome DNA, *J. Parasitol.* **60**:725-727.

Hirai, H., and LoVerde, P.T., 1995, FISH techniques for construction physical maps on schistosome chromosomes, *Parasitol. Today* **11**:310-314.

Hirai, H., and LoVerde, P.T., 1996, Identification of the telomeres on *Schistosoma mansoni* chromosomes by FISH, *J. Parasitol.* **82**:511-512.

Hoffmann, K.F., Johnston, D.A., and Dunne, D.W., 2002, Identification of *Schistosoma mansoni* gender-associated gene transcripts by cDNA microarray profiling, *Genome Biol.* **3**:RESEARCH0041.

Holt, R.A., Subramanian, G.M., Halpern, A., et al., 2002, The genome sequence of the malaria mosquito *Anopheles gambiae, Science* **298**:129-149.

Hu, W., Yan, Q., Shen, D.K., et al., 2003, Evolutionary and biomedical implications of a *Schistosoma japonicum* complementary DNA resource, *Nat. Genet.* **35**:139-147.

Hunter, T.C., Andon, N.L., Koller, A., Yates, J.R., and Haynes, P.A., 2002, The functional proteomics toolbox: methods and applications, *J. Chromatogr. B Analyt. Technol. Biomed. Life Sci.* **782**:165-181.

Johnston, D.A., Blaxter, M.L., Degrave, W.M., Foster, J., Ivens, A.C., and Melville, S.E., 1999, Genomics and the biology of parasites, *Bioessays* **21**:131-147.

Koller, A., Washburn, M.P., Lange, B.M., Andon, N.L., Deciu, C., Haynes, P.A., Hays, L., Schieltz, D., Ulaszek, R., Wei, J., Wolters, D., and Yates, J.R., 3rd, 2002, Proteomic survey of metabolic pathways in rice, *Proc. Natl. Acad. Sci. USA* **99**:11969-11974.

Laudet, V., 1997, Evolution of the nuclear receptor superfamily: early diversification from an ancestral orphan receptor, *J. Mol. Endocrinol.* **19**:207-226.

Le, T.H., Blair, D., and McManus, D.P., 2002, Mitochondrial genomes of parasitic flatworms, *Trends Parasitol.* **18**:206-213.

Le Paslier, M,C., Pierce, R.J., Merlin, F., Hirai, H., Wu, W., Williams, D.L., Johnston, D., LoVerde, P.T., Le Paslier, D., 2000, Construction and characterization of a *Schistosoma mansoni* bacterial artificial chromosome library, *Genomics* **65**:87-94.

Lin, D., Tabb, D.L., and Yates, J.R., 2003, Large-scale protein identification using mass spectrometry, *Biochim. Biophys. Acta* **1646**:1-10.

Link, A.J., Eng, J., Schieltz, D.M., Carmack, E., Mize, G.J., Morris, D.R., Garvik, B.M., and Yates, J.R. 3rd, 1999, Direct analysis of protein complexes using mass spectrometry, *Nat. Biotechnol.* **17**:676-682.

Lockyer, A.E., Olson, P.D., Ostergaard, P., et al., 2003, The phylogeny of the Schistosomatidae based on three genes with emphasis on the interrelationships of *Schistosoma* Weinland 1858, *Parasitology* **126**: 203-224.

McManus, D.P., Le, T.H., and Blair, D., 2004, Genomics of parasitic flatworms, *Int. J. Parasitol.* **34**:153-158.

Merrick, J.M., Osman, A., Tsai, J., Quackenbush, J., LoVerde, P.T., and Lee, N.H., 2003, The *Schistosoma mansoni* gene index: gene discovery and biology by reconstruction and analysis of expressed gene sequences, *J. Parasitol.* **89**:261-269.

Mewes, H.W., Frishman, D., Guldener, U., Mannhaupt, G., Mayer, K., Mokrejs, M., Morgenstern, B., Munsterkotter, M., Rudd, S., and Weil, B., 2002, MIPS: a database for genomes and protein sequences, *Nucleic Acids Res.* **30**:31-34.

Munasinghe, A., Patankar, S., Cook, B.P., Madden, S.L., Martin, R.K., Kyle, D.E., Shoaibi, A., Cummings, L.M., and Wirth, D.F., 2001, Serial analysis of gene expression (SAGE) in *Plasmodium falciparum*: application of the technique to A-T rich genomes. *Mol. Biochem. Parasitol.* **113**:23-34.

Oliveira, G., and Johnston, D.A., 2001, Mining the schistosome DNA sequence database, *Trends Parasitol.* **17**:501-503.

Patankar, S., Munasinghe, A., Shoaibi, A., Cummings, L.M., and Wirth, D.F., 2001, Serial analysis of gene expression in *Plasmodium falciparum* reveals the global expression profile of erythrocytic stages and the presence of anti-sense transcripts in the malarial parasite, *Mol. Biol. Cell.* **12**:3114-3125.

Pearce, E.J., 2003, Progress towards a vaccine for schistosomiasis, *Acta Trop.* **86**:309-313.

Pica-Mattoccia, L., Dias, L.C., Moroni, R., and Cioli, D., 1993, *Schistosoma mansoni*: genetic complementation analysis shows that two independent hycanthone/oxamniquine-resistant strains are mutated in the same gene, *Exp. Parasitol.* **77**:445-449.

Pierrot, C., Godin, C., Liu, J.L., Capron, A., and Khalife, J., 1996, *Schistosoma mansoni* elastase: an immune target regulated during the parasite life-cycle, *Parasitology* **113**:519-526.

Plowman, G.D., Sudarsanam, S., Bingham, J., Whyte, D., and Hunter, T., 1999, The protein kinases of *Caenorhabditis elegans*: a model for signal transduction in multicellular organisms, *Proc. Natl. Acad. Sci. USA* **96**:13603-13610.

Prosdocimi, F., Faria-Campos, A.C., Peixoto, F.C., Pena, S.D., Ortega, J.M., and Franco, G.R., 2002, Clustering of *Schistosoma mansoni* mRNA sequences and analysis of the most transcribed genes: implications in metabolism and biology of different developmental stages, *Mem. Inst. Oswaldo Cruz* **97 Suppl 1**:61-69.

Rabelo, E.M., Franco, G.R., Azevedo, V.A., Pena, H.B., Santos, T.M., Meira, W.S., Rodrigues, N.A., Ortega, J.M., and Pena, S.D., 1997, Update of the gene discovery program in *Schistosoma mansoni* with the expressed sequence tag approach, *Mem. Inst. Oswaldo Cruz* **92**:625-629.

Ramachandran, H., Skelly, P.J., and Shoemaker, C.B., 1996, The *Schistosoma mansoni* epidermal growth factor receptor homologue, SER, has tyrosine kinase activity and is localized in adult muscle, *Mol. Biochem. Parasitol.* **83**:1-10.

Saha, S., Sparks, A.B., Rago, C., Akmaev, V., Wang, C.J., Vogelstein, B., Kinzler, K.W., and Velculescu, V.E., 2002, Using the transcriptome to annotate the genome, *Nat. Biotechnol.* **20**:508-512.

Santos, T.M., Johnston, D.A., Azevedo, V., et al., 1999, Analysis of the gene expression profile of *Schistosoma mansoni* cercariae using the expressed sequence tag approach, *Mol. Biochem. Parasitol.* **103**:79-97.

Short, R.B., and Grossman, A.I., 1981, Conventional giemsa and C-banded karyotypes of *Schistosoma* mansoni and *S. rodhaini*, *J. Parasitol.* **67**:661-671.

Simpson, A.J., Sher, A., and McCutchan, T.F., 1982, The genome of *Schistosoma mansoni*: isolation of DNA, its size, bases and repetitive sequences, *Mol. Biochem. Parasitol.* **6**:125-137.

Skelly, P.J., Da'dara, A., and Harn, D.A., 2003, Suppression of cathepsin B expression in *Schistosoma mansoni* by RNA interference, *Int. J. Parasitol.* **33**:363-369.

Snyder, S.D., and Loker, E.S., Evolutionary relationships among the Schistosomatidae (Platyhelminthes: Digenea) and an Asian origin for *Schistosoma, J. Parasitol.* **86**:283-288.

Sluder, A.E., and Maina, C.V., 2001, Nuclear receptors in nematodes: themes and variations, *Trends Genet.* **17**: 206-213.

Tanaka, M., Hirai, H., LoVerde, P.T., Nagafuchi, S., Franco, G.R., Simpson, A.J., and Pena, S.D., 1995, Yeast artificial chromosome (YAC)-based genome mapping of *Schistosoma mansoni, Mol. Biochem. Parasitol.* **69**:41-51.

The C. elegans Sequencing Consortium, 1998, Genome sequence of the nematode *C. elegans*: a platform for investigating biology, *Science* **282**:2012-2018.

van der Werf, M.J., de Vlas, S.J., Brooker, S., Looman, C.W., Nagelkerke, N.J., Habbema, J.D., Engels, D., 2003, Quantification of clinical morbidity associated with schistosome infection in sub-Saharan Africa, *Acta Trop.* **86**:125-139.

Velculescu, V.E., Zhang, L., Vogelstein, B., and Kinzler, K.W., 1995, Serial analysis of gene expression, *Science* **270**:484-487.

Velculescu, V.E., Zhang, L., Zhou, W., Vogelstein, J., Basrai, M.A., Bassett, D.E. Jr., Hieter, P., Vogelstein, B., and Kinzler, K.W., 1997, Characterization of the yeast transcriptome, *Cell* **88**:243-251.

Velculescu, V.E., Madden, S.L., Zhang, L., et al., 1999, Analysis of human transcriptomes, *Nat. Genet.* **23**:387-388.

Velculescu, V.E., Vogelstein, B., and Kinzler, K.W., 2000, Analysing uncharted transcriptomes with SAGE, *Trends Genet.* **16**:423-425.

Venter, J.C., Smith, H.O., and Hood, L., 1996, A new strategy for genome sequencing, *Nature* **381**:364-366.

Verjovski-Almeida, S., DeMarco, R., Martins, E.A.L., et al., 2003, Transcriptome analysis of the acoelomate human parasite *Schistosoma mansoni, Nat. Genet.* **35**:148-157.

Vermeire, J.J., Osman, A., LoVerde, P.T., and Williams, D.L., 2003, Characterisation of a Rho homologue of *Schistosoma mansoni, Int. J. Parasitol.* **33**:721-731.

Vicogne, J., Pin, J.P., Lardans, V., Capron, M., Noël, C., Dissous, C., 2003, An unusual receptor tyrosine kinase of *Schistosoma mansoni* contains a Venus Flytrap module, *Mol. Biochem. Parasitol.* **126**:51-62.

Washburn, M.P., Wolters, D., and Yates, J.R. 3rd, 2001, Large-scale analysis of the yeast proteome by multidimensional protein identification technology, *Nat. Biotechnol.* **19**:242-247.

Webster, P.J., and Mansour, T.E., 1992, Conserved classes of homeodomains in *Schistosoma mansoni*, an early bilateral metazoan, *Mech. Dev.* **38**:25-32.

Wilson, R.A., and Coulson, P.S., 1998, Why don't we have a schistosomiasis vaccine? *Parasitol. Today* **14**:97-99.

Wippersteg, V., Kapp, K., Kunz, W., and Grevelding, C.G., 2002, Characterisation of the cysteine protease ER60 in transgenic *Schistosoma mansoni* larvae, *Int. J. Parasitol.* **32**: 1219-1224.

Wippersteg, V., Kapp, K., Kunz, W., Jackstadt, W.P., Zahner, H., and Grevelding, C.G., 2002, HSP70-controlled GFP expression in transiently transformed schistosomes, *Mol. Biochem. Parasitol.* **120**:141-150.

Yates, J.R. 3rd, 2000, Mass spectrometry. From genomics to proteomics, *Trends Genet.* **16**:5-8.

Chapter 5

VACCINE DEVELOPMENT
Lessons from the Study of Initial Immune Responses to Invading Larvae

Adrian P. Mountford and Stephen J. Jenkins
Department of Biology, The University of York, York, U.K.

Key words: Vaccine, larvae, innate, skin, Th1

1.   INTRODUCTION

Vaccination against schistosomes has been a long sought after goal of many researchers and the rationale has been well argued in several recent reviews (Wilson and Coulson 1998, Gryseels 2000, Bergquist *et al.* 2002, Todd and Colley 2002). While most vaccination strategies, and nearly all of the current candidate vaccine antigens, have been selected for their ability to reduce the number of challenge worms in immunized animals (anti-larval, or anti-worm), vaccines that restrict the ability of adult worms to produce eggs (anti-fecundity), or reduce pathology (anti-pathology), have also been proposed (James and Colley 2001). This article will focus exclusively upon research aimed at the development of an effective anti-larval vaccine, and will specifically examine recent advances in our understanding of how larval schistosomes first interact with the innate and acquired immune systems. These studies are critical to determining how to present candidate vaccines to the immune system to achieve optimum protection.

1.1   Current Vaccine Candidates

Ten years ago, the World Health Organization selected six of the most promising vaccine candidate antigens to test for their protective efficacy in laboratory mice including glutathione-S-transferase (Sm28), paramyosin (Sm97), triose phosphate isomerase, fatty acid-binding protein (Sm14), internal membrane protein (Sm23), and myosin heavy chain (rIrV5); descriptions of how each was first identified are reviewed by Dunne and Mountford 2001. However, despite trials in two independent laboratories, a target of greater than 40% protection was not achieved for any of the antigens (Bergquist and Colley 1998). This lead to a reappraisal of the vaccine program, with the inclusion of studies to assess human cellular and antibody responses to the candidate antigens (Bergquist et al. 2002). In parallel, *S. haematobium* GST (Bilhvax) has been taken through Phase I and II clinical trials and now awaits testing in larger Phase III trials. In addition to the six antigens tested above, at least another 10 protective antigens have been reported (Dunne and Mountford 2001) and there are probably tens/hundreds more recombinant expressed proteins stored in various freezers around the world that remain untested (or unreported) for their protective capacity. Despite progress made with ShGST, and the plethora of prospective vaccine candidates, an effective vaccine for use against all three schistosome species in humans appears still to be some way off.

1.2   Why don't we have a vaccine? Unresolved issues

If we take a step back from these promising developments, there remain three long-standing important but controversial questions about the overall strategy of vaccine development against schistosomiasis:
1). Do we have the correct antigens?
2). Do we know how to formulate them in such a way that they are processed and presented by the immune system leading to the induction of protective immunity?
3). Do we know what type of immune response is protective?

Leaving the first question aside for a different forum, the following treatise focuses upon research of particular relevance to resolving questions 2 and 3. Defining the mechanism of protective immunity has lead to a debate about the extent and nature of protective immunity in humans but also whether protective mechanisms observed to occur in experimental hosts are relevant to the human population (Gryseels 2000, James and Colley 2001, Druille et al. 2002, Bergquist et al. 2002). Arguments have been made for the protective effects of both IFNγ (Th1-associated) and antibody (Th2-associated) mediated immune mechanisms, although the debate is complicated because it is difficult to examine immune responses that are exclusively induced by larval stages in humans. The absence of agreement on the nature of the protective response perhaps explains the lack of coherence in the choice of adjuvant/delivery system for use with candidate vaccine antigens. However, our relative ignorance of the innate immune response to invading larvae is another significant factor. In this respect, many adjuvants act directly upon the innate immune system, either through the inclusion of molecules of bacterial origin (*e.g.* lipid A derivatives), or by causing the release of endogenous 'danger' signals (*e.g.* saponin derivatives) (Schijns 2000).

1.3   The innate / acquired immune response interface

Since Janeway first proposed the intimate linkage of the innate and acquired immune systems (Medzitov and Janeway 1997), it has become popular to regard the initial innate immune recognition of pathogenic agents as a prerequisite and deciding factor in the development of the ensuing acquired immune response.   A central point of their proposal is that pathogenic organisms possess a number of conserved molecular motifs (pathogen-associated molecular patterns; PAMPs) which are essential for the survival of the microbe but against which the host has evolved a limited number of receptors (pattern-recognition receptors; PRRs).   The expression of PRRs is confined largely to accessory cells of the innate immune system, some of which have antigen processing and presenting function (*e.g.* dendritic cells (DCs) and macrophages).   The hypothesis follows that innate immune cells at the site of initial antigen contact are important constituents in the subsequent development of the acquired immune response.   In this context, the skin is where initial host immune recognition of schistosome larvae occurs.   This site is rich in numerous immuno-competent cell types including antigen presenting cells (APC) such as dermal DCs, epidermal Langerhan's cells (LC) and macrophages.   These skin-derived APCs are thought to be important in the priming of T cell responses in the draining lymphoid tissue (sdLN). Consequently, analysis of innate immune responses to schistosomes is an area of increasing interest, and events that occur in the skin are critical to our understanding of how protective immunity is initiated and how the parasite circumvents the host's response.

2.   LESSONS FROM THE STUDY OF INITIAL IMMUNE RESPONSES TO INVADING LARVAE

2.1   Protection induced by the Radiation-Attenuated Vaccine: The gold standard?

The radiation-attenuated (RA) schistosome vaccine model stands out for its ability to induce consistently high levels of protective immunity in a wide range of experimental hosts.   It is our contention that the study of immune mechanisms elicited by the RA vaccine in the mouse is an important model that will provide insights into how protective immune responses against larval schistosomes are induced. Specifically, investigations of the stimulation of innate and early acquired immune responses by RA larvae set an essential context for the rational design of novel adjuvants and delivery systems. In this model, RA larvae are generally considered to stimulate protective immunity dominated by Th1-type cell-mediated responses (Street et al. 1999), although antibodies (usually Th2-associated) also have a role in certain circumstances (Wynn and Hoffmann 2000, Mountford et al. 2001).

The exposure of cercariae to optimal doses of ionizing radiation has proved to be the most reliable way of inducing high levels of protective immunity (Coulson 1997). Most studies have examined RA *S. mansoni*, but the approach has also been applied to *S. japonicum, S. haematobium* and *S. bovis* (Dunne and Mountford 2001). High levels of protection have been observed in a number of experimental models for *S. mansoni* ranging from the mouse, the brown rat, the guinea pig, and non-human primates such as the olive baboon and most recently, the chimpanzee (Eberl et al. 2001). Nevertheless, studies of protection in the mouse model have been the most intensive largely because of its permissiveness to infection, its well-characterized physiology and genetics, the wide availability of immunological reagents, and the opportunity to experimentally manipulate the immune system (*e.g.* using mice transgenic for immune response genes).

2.1.1   Normal versus RA schistosome larvae; comparison of their ability to migrate and induce protection

In the mouse, 60-70% protection can be induced by a single exposure to optimally attenuated *S. mansoni* larvae. In contrast, normal parasites are not thought to elicit significant levels of acquired immunity in the absence of egg-induced pathology (Wilson, 1990). A major focus of research has been to identify features of the RA vaccine that may explain why it is so effective while exposure to normal parasites is not. One early observation showed that RA larvae were much slower to migrate from the skin than normal ones (Fig. 5-1; Mountford et al. 1988). Consequently, the persistence of RA live parasites in this site represents a substantial antigenic load and stimulus to the dermal immune environment. The study also revealed that a substantial number of both normal and RA larvae could be detected in the skin-draining lymph nodes (sdLN). Interestingly, while normal larvae reached peak numbers in the sdLN on day 4-5, RA larvae arrived slightly later and persisted for many days. Together with free parasite antigen draining from the skin, and that carried to the sdLN by migrating APCs, this represents a much greater load of parasite material being delivered to the highly active immune priming environment of the sdLN. Finally, while normal parasites migrated via the lungs en-route to the hepatic portal system, RA larvae had a truncated migration progressing no further than the lungs where they expire. Why RA larvae suffer from a delayed and truncated migration is not yet clear, although it is possible that radiation-induced damage to their neuro-muscular co-ordination impairs their ability to negotiate difficult tissue sites such as the skin, sdLN and lungs (Harrop and Wilson 1993).

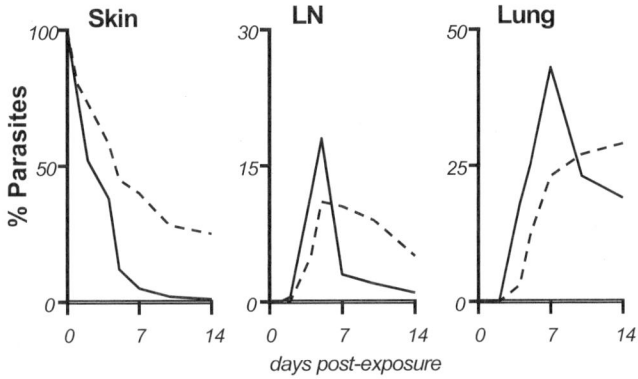

Figure 5-1. Distribution of 75-Se-labelled parasites in mouse tissues. Normal larvae (solid line) and RA larvae (dashed line). Redrawn from Mountford et al. 1988.

2.1.2   Priming of innate immune responses in the skin

Since the skin is the initial site of exposure to either normal, or immunizing RA larvae, it is logical to assume that immune events in this site will have an important impact on the outcome of the ensuing adaptive immune response. The schistosome larva is a large target and unlikely to be killed by innate immune defense mechanisms. However, its entry into the skin will present the host accessory cells with a number of molecules with PAMP-like features. The parasite's ability to release various proteolytic enzymes (to facilitate its migration through the skin layers) may cause tissue destruction alerting the host to 'danger', which is also a key trigger for cells of the innate immune response (Galucci *et al.* 1999).

Vaccine development

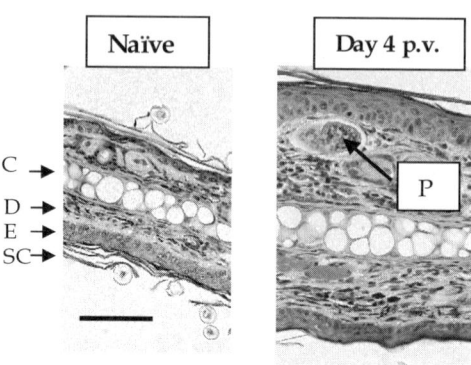

Figure 5-2. Micrographs showing cellular influx after exposure on the mouse pinna to RA larvae. SC, stratum corneum; E, epidermis; D, dermis; C, cartilage; P, parasite. Bar = 50μm. Photograph courtesy of K.G. Hogg.

In vivo, schistosome larvae quickly negotiate the stratum corneum and enter the epidermis. The invading parasites elicit rapid inflammation of the skin site evident within 6-24 hrs. This increase in skin thickness is caused by oedema and by marked focal influx of mononuclear and poly-morphonuclear and cells (Fig. 5-2).

Immunohistochemical analysis of the skin following exposure to RA larvae shows an abundance of cells positive for the neutrophil marker 7/4, but also for myeloid markers such as CD11b and CD11c, which peak in number by day 4 within the dermis (Riengrojpitak et al. 1998, Hogg et al. 2003a). Cells with putative APC function are an important feature of the skin in mice exposed to the RA vaccine and antigen-laden cells that emigrate form a crucial conduit of parasite material to the sdLN. A large number of cells which spontaneously migrate from *in vitro*-cultured skin biopsies of vaccinated mice are MHCII$^+$, consistent with them being APCs. A population of IL-12p40$^+$ cells are also detected in the emigrant population which are CD11b$^+$, CD11c$^+$ and/or F4/80$^+$ indicating they are DCs and/or macrophages (Hogg et al. 2003a). In this context, recent interest in our laboratory has focused upon the role of LCs following vaccination. Their position in the epidermis makes them well placed to take up antigen in the first two days after vaccination before larvae breach the epidermal/dermal basement membrane. Using a specific antibody against langerin (CD207), we have found that LCs quickly migrate from the epidermis and increased numbers can be detected in the dermal emigrant population and in the sdLN (Kumkate, Hogg and Mountford, unpublished data).

2.1.3   Pro-inflammatory immune responses in the skin

The cellular influx in the skin is undoubtedly initiated by resident cells that, upon activation by invading larvae, secrete a cascade of chemokines and cytokines to recruit further cells. Analysis of the cytokine environment following vaccination is critical to understanding the traffic and priming of appropriate immune cells that will guide the development of the acquired immune response. The earliest molecules to be detected are three CC chemokines MIP-1α, MIP-1β and eotaxin released at peak levels within 24 hours of parasite exposure (Hogg et al. 2003a). MIP-1α and MIP-1β are attractants for monocytes, lymphocytes and immature DCs into the skin, and both are linked to the development of Th1-type responses (Dorner et al. 2002). However, these chemokines appear to be the first step of an inflammatory cascade involving the transient production of IL-1β and IL-6 that can lead to the migration of LCs out of the skin (Wang et al. 1999). Such cytokines are also important in activating APCs by increasing their expression of MHC class II and aid the production of other cytokines such as IL-12 and IL-18.

One of our recent findings is that the skin is an important site for the production of IL-12 following vaccination (Fig. 5-3; Hogg et al. 2003a). Levels of this cytokine increased rapidly following vaccination and are sustained out to at least day 14. Since IL-12 is a potent cytokine involved in the differentiation of $CD4^+$ cells towards the Th1 type (Trinchieri, 2003), its production in the skin is clearly conducive to the promotion of protective Th1-type responses in this vaccine model. Indeed, vaccinated IL-12p40-deficient mice develop polarized Th2-type immune responses (Hogg et al. 2003b) that are not highly protective (Anderson et al. 1998). Moreover, in vaccinated IL-4R deficient mice, levels of IL-12p40 in the skin are even greater indicating some form of IL-4/IL-13 mediated regulation of IL-12p40 production (Mountford et al. 2001). In contrast, normal parasites induce only a transient production of IL-12p40 correlating with their rapid exit from the skin (Fig. 5-3; Hogg et al. 2003a). Consequently, we have concluded that sustained IL-12p40 in the skin is an important feature in vaccinated mice due to the persistence of live parasites (releasing antigenic material) in the skin. Although another Th1-inducing cytokine, IL-18 was also detected in the supernatants of in vitro-cultured skin biopsies, its production was not diminished in IL-12p40-deficient animals (Hogg et al. 2003b) and similar amounts were detected for infected and vaccinated mice (Hogg et al. 2003a) indicating it does not have a critical role in this vaccine model.

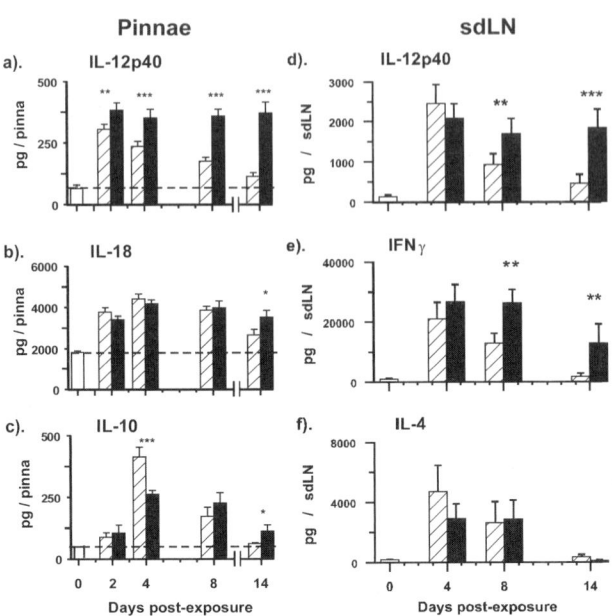

Figure 5-3. Cytokine production in the pinnae and sdLN following exposure to normal (hatched) or RA (solid) larvae. Production of IL-12p40, IL-18 and IL-10 after 18 hr in vitro culture in the absence of added antigen. Production of IFNγ and IL-4 after 72 hr culture in response to parasite antigen stimulation. Significance values are for infected vs vaccinated groups at each time points (Reprinted with permission from Hogg et al. 2003a).

2.1.4   Anti-inflammatory responses in the skin

Alongside the ability of both normal and RA schistosomes to induce inflammation of the skin, it has been shown that normal larvae also have an anti-inflammatory effect. Studies by Ramaswamy et al. have linked the production by keratinocytes of the potent IL-1 antagonist 'IL-1ra' to stimulation by the excretory/secretory products of transforming cercariae (Ramaswamy et al. 1995). Normal parasites are also efficient stimulators of IL-10 production in mouse skin (Fig. 5-3: Angeli et al. 2001, Hogg et al. 2003a) and by human keratinocytes (Ramaswamy et al. 2000), and it is proposed that induction of this cytokine is dependent upon the secretion of parasite-derived prostaglandin E_2 (PGE_2) upon reaction with free fatty acids in the skin (Ramaswamy et al. 2000). The induction of IL-10 is significant because it regulates the production of proinflammatory cytokines (e.g. IL-1β and IL-12), and the expression of MHC class II and costimulatory molecules (CD80 and CD86) on potential APCs. In this context, Ramaswamy et al. (2000) recorded that RA schistosome larvae were very poor inducers of IL-10, leading to the conclusion that the ability of RA larvae to efficiently prime for protective immune responses was linked to their inability to stimulate the production of IL-10. However, we have found that RA larvae also stimulate IL-10 production (Fig. 5-3) and in its absence, increased levels of IL-1β, IL-12p40 from in vitro-cultured skin biopsies were associated with elevated levels of inflammation (Hogg et al. 2003b). The absence of IL-10 also led to a greater number of Ia$^+$ cells in the dermis and an increased number of MHC II$^+$ and CD86$^+$ cells in the population of dermal emigrant cells. Consequently, while the mechanism for IL-10 induction via host, or parasite-derived eicosanoids remains valid, it is likely that it is common to both normal and RA larvae although the kinetics of induction may differ between the two forms of larvae.

The role of eicosanoids in the early innate immune response to schistosomes is of further significance since it has been reported that the vaccine candidate Sm28 GST has PGD$_2$ synthase activity (Herve et al. 2003). Initially, it was found that the migration of LCs from the epidermis was inhibited in mice exposed to normal cercariae (Angeli et al. 2001). This was caused by the production of parasite-derived PGD$_2$ (but not PGE$_2$) which, through the activation of the D prostanoid receptor (DP), interfered with TNFα-induced LC migration. However, in contrast to PGE$_2$, a role for IL-10 was not found although this cytokine is thought to specifically impede LC migration (Wang et al. 1999). Recently, Sm28 GST was found to abrogate the emigration of LCs but not in mice deficient for DP (Herve et al. 2003). Consequently, it appears that Sm28 GST actively aids the production of PGD$_2$ and is a homologue of mammalian hematopoietic PGD synthase. It is not known whether there is a difference in the ability of RA compared to normal larvae to secrete GST, or produce PGD$_2$. If RA larvae are found to be deficient in GST, or PGD$_2$, as suggested for PGE$_2$ (Ramaswamy et al. 2000), it is conceivable that the greater levels of certain pro-inflammatory cytokines in response to RA larvae may be a reflection of a radiation-induced biochemical change in the immunizing larvae.

Schistosome larvae may possess specific apoptotic factors that cause the death of immune cells. Chen *et al.* (2002) reported that skin-stage parasites release a molecule that up-regulates FasL and Fas on CD4$^+$ and CD8$^+$ cells and caused enhanced levels of apoptosis. The killing of cells by normal larvae has been suggested to act as a parasite survival strategy. However, the authors also suggest that the molecular component is absent in RA larvae thus potentially explaining why these parasites are more efficient at stimulating immune responses. This is an attractive notion, but it implies that irradiation somehow eliminates a protein that would be normally pre-formed in cercariae prior to irradiation. Further research into this intriguing phenomenon is eagerly awaited.

2.1.5   Evidence for the existence of schistosome PAMPs in priming innate responses

The role of schistosome-derived PGE_2 and PGD_2 outlined above highlights the impact that parasite molecules are likely to have on cells of the innate immune system, most obviously at the site of penetration. A further key area of research is likely to center upon the definition of schistosome molecules which have PAMP-like activity by ligating host PRRs and stimulating pro-inflammatory reactions. Identification of such molecules would aid our understanding of how the invading larvae triggers the innate immune response and may pave the way to the identification of novel immunomodulators, or adjuvants.

Since schistosome cercariae originate from a non-sterile aquatic environment, it is possible that the cytokine cascade observed to occur in the skin of vaccinated mice might actually be caused by bacterial contaminants that could enter the host following cercarial penetration. As such, lipopolysaccharide (LPS) from gram-negative bacteria binding to Toll-like receptor 4 (TLR-4) is a major potential cause of IL-12 production by macrophages and DCs (Aikira et al. 2001). However, we found that the production of IL-12p40 by skin biopsies from vaccinated TLR-4 gene deficient C3H/HeJ mice was not significantly lower and protective immunity was not reduced (Hogg et al. 2003a). Although contaminating ligands for other PRRs may enter the skin during vaccination, it seems likely that the parasite possesses its own PAMP-like molecules. Indeed, we have identified excretory/secretory molecules released during cercarial transformation as potent stimulators of cytokine release (including IL-6 and IL-12) by in vitro cultured peritoneal exudate macrophages and bone marrow-derived DCs (Jenkins and Mountford, manuscript in preparation). The identification of TLRs, or PRRs involved in the initiation of the innate immune response is therefore a key area for investigation and would be beneficial to our understanding of how schistosome larvae stimulate the innate immune response. Incorporation of putative schistosome PAMPs into vaccine constructs could help the induction of protective immune responses.

2.2   Priming of acquired immune responses in the skin-draining LNs

Activation of innate immune cells in the skin is an important precursor to the subsequent priming of acquired responses in the sdLN. Some of the responsive cells in the skin will be APCs that take up parasite antigen and migrate to the T cell areas of the sdLN where they can present antigen. Indeed, as stated earlier, CD11c[+] and F4/80[+] skin emigrant cells capable of secreting IL-12p40 are a feature of mice exposed to RA larvae (Hogg et al. 2003a). It is not known whether these cells are epidermal-derived LCs, or dermal DCs and macrophages, but we have established a 2-3 fold increase in the numbers of CD207$^+$ LCs in the sdLN of vaccinated mice (Kumkate, Hogg and Mountford, unpublished data). Antigen released by parasites in the skin may also drain to the sdLN, whilst at least 15% of RA larvae deliver themselves to the sdLN (Fig. 5-1). In both cases, parasite antigen would be processed and presented to T cells *in situ*. Priming of immune responses in the sdLN is important since routes of delivery which by-pass this site, or where the sdLN have been removed, lead to a reduction in the efficacy of the vaccine (Mountford and Wilson 1990). Analysis of the lymphocyte and cytokine responses in the sdLN reveals that there is little activity up to day 2 to 3 post-vaccination but significant production of IL-12p40 occurs in both infected and vaccinated mice by day 4 (Fig. 5-3; Hogg et al. 2003a). However, while this is maintained in mice exposed to the RA vaccine, levels of IL-12p40 rapidly decline in infected mice.

This pattern of cytokine production is reflected in the production of antigen-specific IFNγ. By day 14, IFNγ was sustained in the vaccinated group only, demonstrating a strong bias towards a Th1-dominated immune environment (Hogg et al. 2003a). This bias is dependent upon the presence of IL-12 and IL-10 since the absence of IL-12p40 leads to a polarized Th2-type environment in the sdLN (Anderson et al. 1998, Hogg et al. 2003b), whilst the absence of IL-10 promotes a polarized Th1-type environment (Hoffmann et al. 2000, Hogg et al. 2003b). In the absence of IL-10, greater numbers of MHC II$^+$ APCs are evident in the sdLN, and there are more CD4 cells positive for CD25, CD28 and CTLA-4, indicative of a more active immune priming environment (Hogg et al. 2003b). This is reflected by contrasting levels of resistance that are substantially reduced in the absence of IL-12, but enhanced in IL-12-rich environments. Indeed, administration of recombinant IL-12 to vaccinated mice can elevate the protection towards the ultimate goal of sterile immunity (Wynn et al. 1996, Anderson et al. 1998). However, recently it has become clearer that protective immunity can occur in the absence of strong Th1-type responses (Anderson et al. 1998, Hoffmann et al. 2000), leading to the conclusion that Th2-associated antibody-mediated immune mechanisms are also protective, even after exposure to a single vaccine (Mountford et al. 2001). Indeed, both Th subsets may need to be activated for optimum levels of protection (Wynn and Hoffmann 2000).

The priming of CD4$^+$ cells in the sdLN is a vital component of the success of the RA vaccine and the induction of biased Th1-type responses are an important protective mechanism. Apart from manipulating the IL-12 / IL-10 cytokine environment, little attention has been paid to other components of immune priming which may boost Th1-type responses. Of immediate interest are the IL-12-like cytokines IL-23 and IL-27 (Brombacher et al. 2003). Both have important, but distinct, roles in aiding the development of Th1-type responses. Specifically, IL-23 from keratinocytes is a key cytokine in the initiation of cutaneous responses (Kopp et al. 2003).

The role of costimulatory molecules has also been largely ignored. In this respect, levels of CD154 expression on cells in the sdLN increase after vaccination, while the administration of anti-CD40 antibody to vaccinated mice causes an increase in the bias of the immune response towards the Th1 pole (Hewitson and Mountford, unpublished data). As such, stimulation of the CD154/CD40 pathway is of likely importance. It remains to be determined whether other costimulatory molecules, or cofactors, might have a significant pro-, or anti-, inflammatory effect such that APCs are activated to promote the induction of Th1 as opposed to Th2-type responses. The description of several new members of the B7 family (Coyle et al. 2001) opens the possibility that they might have a regulatory effect early after vaccination and their manipulation may prove beneficial.

2.3   Recruitment of primed immune cells to the lungs; the final step

In order to achieve optimum levels of protection, recruitment of antigen-primed CD4$^+$ cells to the lungs appears to be an important feature of the RA vaccine model (Coulson 1997). If RA parasites do not reach the lungs, or if priming of the lungs occurs without prior stimulation in the sdLN, a protective response fails to be generated (Mountford et al. 1992). Moreover, in this vaccine model, the lungs form the site where the majority of challenge parasite elimination occurs via an IFNγ and TNF-dependent cell-mediated immune response (Street et al. 1999). The immune events that occur in the lung following priming and their relationship to the generation of immune effector responses have been reviewed extensively elsewhere (Coulson, 1997, Dunne and Mountford, 2001) and will not be described here. However, it remains a challenging proposition to the architect of a future vaccine to ensure that an adequate population of primed lymphocytes inhabits the pulmonary tissue where they can await the arrival of challenge parasites. Whether immune priming in one site (*i.e.* the sdLN) can be engineered to cause the spillover of sufficient cells into the lungs remains to be established. Perhaps multiple exposures to candidate vaccines will be necessary but in this context antibody-mediated protective immune mechanisms may also operate to confer protection (Wynn and Hoffman,

2000). As yet, it is not known against which stage of parasite antibody-mediated defense mechanisms operate.

In conclusion, the analysis of early innate and acquired immune responses to invading larvae provides valuable insights into how we might trigger responses that are host protective. However, it is also evident that the parasite has evolved a survival strategy to successfully exit the skin and avoid the host's attention. Further work, particularly on the identification of schistosome PAMPs and how they interact with host accessory cells will benefit the design of appropriate vaccine adjuvants and immunomodulators. It is to be expected that both pro-inflammatory and anti-inflammatory agents will be identified, the balance of which will be essential to development of anti-parasite vaccination strategies. As such, the RA vaccine remains a valuable research model from which important lessons can be learnt about the nature of protective immunity and how it is stimulated.

ACKNOWLEDGEMENTS

I would like to thank past and present members of my research group, Karen G. Hogg, Supeecha Kumkate and James P. Hewitson who have contributed towards some of the research described in the above article. A.P.M. was in receipt of a Wellcome Trust University fellowship (Grant # 056213.). Financial support was also provided by the BBSRC (S.J.J. & J.P.H.) and the Royal Thai government (S.K.).

REFERENCES

Akira, S., K. Takeda, and T. Kaisho, 2001, Toll-like receptors: critical proteins linking innate and acquired immunity, *Nat. Immunol.* **2:**675-680.

Anderson, S., V.L. Shires, R.A. Wilson, and A.P. Mountford, 1998, In the absence of IL-12, the induction of Th1-mediated protective immunity by the attenuated schistosome vaccine is impaired, revealing an alternative pathway with Th2-type characteristics, *Eur. J. Immunol.* **28:**2827-2838.

Angeli, V., C. Faveeuw, O. Roye, J. Fontaine, E. Teissier, A. Capron, I. Wolowczuk, M. Capron, and F. Trottein, 2001, Role of the parasite-derived prostaglandin D2 in the inhibition of epidermal Langerhans cell migration during schistosomiasis infection, *J. Exp. Med.* **193:**1135-1147.

Bergquist, R., M. Al-Sherbiny, R. Barakat, and R. Olds, 2002, Blueprint for schistosomiasis vaccine development, *Acta Trop.* **82:**183-192.

Bergquist, N.R. and D.G. Colley, 1998, Schistosomiasis vaccines: research to development. *Parasitol. Today* **14:**99-104.

Brombacher, F., R.A. Kastelein, and G. Alber, 2003, Novel IL-12 family members shed light on the orchestration of Th1 responses, *Trends Immunol.* **24:**207-212.

Chen, L., K.V. Rao, Y.X. He, and K. Ramaswamy, 2002, Skin-stage schistosomula of *Schistosoma mansoni* produce an apoptosis-inducing factor that can cause apoptosis of T cells. *J. Biol. Chem.* **277**:34329-34335.

Coulson, P.S., 1997, The radiation-attenuated vaccine against schistosomes in animal models: paradigm for a human vaccine? *Adv. Parasitol.* **39**:271-336.

Coyle, A.J., and J.C. Gutierrez-Ramos, 2001, The expanding B7 superfamily: increasing complexity in costimulatory signals regulating T cell function, *Nat. Immunol.* **2**:203-209.

Dorner, B.G., A. Scheffold, M.S. Rolph, M.B. Huser, S.H. Kaufmann, A. Radbruch, I.E. Flesch, and R.A. Kroczek, 2002, MIP-1alpha, MIP-1beta, RANTES, and ATAC/lymphotactin function together with IFN-gamma as type 1 cytokines, *Proc. Natl. Acad. Sci. USA* **99**:6181-6186.

Druilhe, P., P. Hagan, and G.A. Rook, 2002, The importance of models of infection in the study of disease resistance, *Trends Microbiol.* **10**:S38-S46.

Dunne, D.W. and A.P. Mountford, 2001, Resistance to infection in humans and animal models, in: *Schistosomiasis,* A.A.F. Mahmoud, ed. Imperial College Press, pp. 133-121.

Eberl, M., J.A. Langermans, P.A. Frost, R.A. Vervenne, G.J. van Dam, A.M. Deelder, A.W. Thomas, P.S. Coulson, and R.A. Wilson, 2001, Cellular and humoral immune responses and protection against schistosomes induced by a radiation-attenuated vaccine in chimpanzees, *Infect. Immun.* **69**:5352-5362.

Gallucci, S., M. Lolkema, and P. Matzinger, 1999, Natural adjuvants: endogenous activators of dendritic cells, *Nat. Med.* **5**:1249-1255.

Gryseels, B., 2000, Schistosomiasis vaccines: a devils' advocate view. *Parasitol. Today* **16**:46-48.

Harrop, R., and R.A. Wilson, 1993, Irradiation of *Schistosoma mansoni* cercariae impairs neuromuscular function in developing schistosomula, *J. Parasitol.* **79**:286-289.

Herve, M., V. Angeli, E. Pinzar, R. Wintjens, C. Faveeuw, S. Narumiya, A. Capron, Y. Urade, M. Capron, G. Riveau, and F. Trottein, 2003, Pivotal roles of the parasite PGD2 synthase and of the host D prostanoid receptor 1 in schistosome immune evasion, *Eur. J. Immunol.* **33**:2764-2772.

Hoffmann, K.F., S.L. James, A.W. Cheever, and T.A. Wynn, 1999, Studies with double cytokine-deficient mice reveal that highly polarized Th1- and Th2-type cytokine and antibody responses contribute equally to vaccine-induced immunity to *Schistosoma mansoni, J. Immunol.* **163**:927-938.

Hogg, K.G., S. Kumkate, S. Anderson, and A.P. Mountford, 2003a, Interleukin-12 p40 Secretion by Cutaneous CD11c(+) and F4/80(+) Cells Is a Major Feature of the Innate Immune Response in Mice That Develop Th1-Mediated Protective Immunity to *Schistosoma mansoni, Infect. Immun.* **71**:3563-3571.

Hogg, K.G., S. Kumkate, and A.P. Mountford, 2003b, IL-10 regulates early IL-12-mediated immune responses induced by the radiation-attenuated schistosome vaccine, *Int. Immunol.* **15**:1451-1459.

James, S.L. and D.G. Colley, 2001, Progress in vaccine development, in: *Schistosomiasis,* A.A.F. Mahmoud, ed. Imperial College Press, pp. 469-497.

Kopp, T., P. Lenz, C. Bello-Fernandez, R.A. Kastelein, T.S. Kupper, and G. Stingl, 2003, IL-23 production by cosecretion of endogenous p19 and transgenic p40 in keratin 14/p40 transgenic mice: evidence for enhanced cutaneous immunity, *J. Immunol.* **170**:5438-5444.

Medzhitov, R. and C.A. Janeway, 1997, Innate immunity: impact upon the adaptive immune response, *Curr. Opin. Immunol.* **9**:4-9.

Mountford, A.P., P.S. Coulson, and R.A. Wilson, 1988, Antigen localization and the induction of resistance in mice vaccinated with irradiated cercariae of *Schistosoma mansoni*, *Parasitology* **97**:11-25.

Mountford, A.P., and R.A. Wilson, 1990, *Schistosoma mansoni*: the effect of regional lymphadenectomy on the level of protection induced in mice by radiation-attenuated cercariae, *Exp. Parasitol.* **71**:463-469.

Mountford, A.P., P.S. Coulson, R.M. Pemberton, L.E. Smythies, and R.A. Wilson, 1992, The generation of interferon-gamma-producing T lymphocytes in skin-draining lymph nodes, and their recruitment to the lungs, is associated with protective immunity to *Schistosoma mansoni*, *Immunology* **75**:250-256.

Mountford, A.P., K.G. Hogg, P.S. Coulson, and F. Brombacher, 2001, Signaling via interleukin-4 receptor alpha chain is required for successful vaccination against schistosomiasis in BALB/c mice, *Infect. Immun.* **69**:228-236.

Ramaswamy, K., B. Salafsky, S. Potluri, Y. X. He, J.W. Li, and T. Shibuya, 1995, Secretion of an anti-inflammatory, immunomodulatory factor by Schistosomulae of *Schistosoma mansoni*, *J. Inflamm.* **46**:13-22.

Ramaswamy, K., P. Kumar, and Y.X. He, 2000, A role for parasite-induced PGE2 in IL-10-mediated host immunoregulation by skin stage schistosomula of *Schistosoma mansoni*, *J. Immunol.* **165**:4567-4574.

Riengrojpitak, S., S. Anderson, and R.A. Wilson, 1998, Induction of immunity to *Schistosoma mansoni*: interaction of schistosomula with accessory leucocytes in murine skin and draining lymph nodes, *Parasitology* **117**:301-309.

Schijns, V.E., 2000, Immunological concepts of vaccine adjuvant activity, Curr. Opin. Immunol. **12**:456-463.

Street, M., P.S. Coulson, C. Sadler, L.J. Warnock, D. McLaughlin, H. Bluethmann, and R.A. Wilson, 1999, TNF is essential for the cell-mediated protective immunity induced by the radiation-attenuated schistosome vaccine, *J. Immunol.* **16**:4489-4494.

Todd, C.W., and D.G. Colley, 2002, Practical and ethical issues in the development of a vaccine against schistosomiasis mansoni, *Am. J. Trop. Med. Hyg.* **66**:348-358.

Trinchieri, G., 2003, Interleukin-12 and the regulation of innate resistance and adaptive immunity, *Nat. Rev. Immunol.* **3**:133-146.

Wang, B., P. Amerio, and D.N. Sauder, 1999, Role of cytokines in epidermal Langerhans cell migration, *J. Leukoc. Biol.* **66**:33-39.

Wang, B., L. Zhuang, H. Fujisawa, G.A. Shinder, C. Feliciani, G.M. Shivji, H. Suzuki, P. Amerio, P. Toto, and D.N. Sauder, 1999, Enhanced epidermal Langerhans cell migration in IL-10 knockout mice, *J. Immunol.* **162**:277-283.

Wilson, R. A., 1990, Leaky livers, portal shunting and immunity to schistosomes, *Parasitol. Today* **6**:354-358.

Wilson, R.A. and P.S. Coulson, 1998, Why don't we have a schistosomiasis vaccine? *Parasitol. Today* **14**:97-99.

Wynn, T.A., A. Reynolds, S. James, A.W. Cheever, P. Caspar, S. Hieny, D. Jankovic, M. Strand, and A. Sher, 1996, IL-12 enhances vaccine-induced immunity to schistosomes by augmenting both humoral and cell-mediated immune responses against the parasite, *J. Immunol.* **157**:4068-4078.

Wynn, T.A., and K.F. Hoffmann, 2000, Defining a schistosomiasis vaccination strategy - is it really Th1 versus Th2? *Parasitol. Today* **16**:497-501.

Chapter 6

THE SURFACE OF SCHISTOSOMES WITHIN THE VERTEBRATE HOST

Patrick J. Skelly
Department of Biomedical Sciences, Tufts University School of Veterinary Medicine, Grafton, MA 01536

Key words: Tegument, nutrient transport, phosphatase, esterase, receptor

1.   INTRODUCTION

The schistosome surface has been the subject of considerable research. It is the site of intimate host/parasite interaction; nutrients are taken up across the body surface of the parasite and environmental sensing occurs there. Great interest exists in identifying and characterizing the molecules that comprise the host-exposed parasite surface for two main reasons: first, in order to gain a better understanding of surface biochemistry and cell biology and second, since molecules exposed at the parasite surface should be accessible to host immune effectors and therefore could be used as vaccine targets to evoke protective immunity. Furthermore, interest in the parasite surface is increased because it has long been known to have host molecules associated with it. The surface is therefore a chimera of host and parasite origin. This paper focuses on schistosome life stages that reside within the vertebrate host. The host-exposed surface molecules of parasite origin are first described and later, host molecules that have been reported to be linked to the parasite surface are examined.

2.   THE TEGUMENT

The major interface between the schistosome and its external environment is called the tegument. The molecules that comprise the host-interactive surface of this tegument are the focus of this paper. This surface is unusual since it comprises not one but two lipid bilayers [1]. This double-bilayered (or heptalaminate) outer membrane is unique to blood-dwelling trematodes such as schistosomes and is not found in trematode parasites occupying other habitats [2]. The outer membrane set of schistosomes is specific for intra-vertebrate stage parasites and forms soon after the infectious cercariae invade the final host. Beneath the tegument surface lies a cytoplasmic layer containing mitochondria and two different kinds of secretory bodies commonly called discoid bodies and multilamellar vesicles [3]. These secretory bodies are thought to be the precursors of the outer tegumental membranes. The tegument lacks lateral membranes. This means that the tegumental cytoplasm extends as a continuous unit or syncytium around the entire body [4]. The cytoplasm is connected by numerous, thin cytoplasmic processes to cell bodies or cytons that lie beneath the peripheral muscle layers. These contain nuclei, endoplasmic reticula, Golgi complexes and mitochondria. Cytons actively synthesize the secretory bodies which move along the cytoplasmic connections to the outer cytoplasmic layer [4]. Following cercarial infection, the cercarial outer membrane is cast off and secretory bodies migrate from the cytons to the parasite surface and fuse there [5]. This results in an initial patchy, multi-lipid-bilayered covering. These lipid deposits resolve over several hours to form the final apical, double-bilayer typical of the intravascular forms [5,6].

Here we examine what is known about the molecules that are exposed on the outer tegumental membrane at this host-parasite interface.

3.   PARASITE SURFACE PROTEINS

A number of parasite proteins have been identified in the outer tegument of several schistosome species. Such proteins belong to a variety of classes and include those involved in nutrient uptake by the worms, as well as enzymes and receptors and several proteins of unknown function. Analysis of the *S. mansoni* transcriptome has resulted in the identification of several new, potentially surface-exposed molecules [7].

Despite the fact that adult schistosomes have a mouth and gut, many nutrients are imported into the body across the tegument surface. Surface transporter proteins for the import of glucose (SGTP4) and selected amino acids (SPRM1lc) have been characterized [6,8-11]. An array of enzyme activities have been detected at the surface and, in several cases, the enzymes involved have been identified for instance, alkaline phosphatase [12-16] and acetylcholinesterase [14,17-20]. A variety of receptors or receptor homologs have been characterized including a collection of receptors for members of the host's complement cascade such as C2 [21], C3 [22], C8 and C9 [23]. Several structural proteins or proteins of unknown function have also been localized at the parasite surface and these include the antigen Sm23 [24-26] and a 200 kDa glycoprotein, Sm200 [27,28]. Tables 1 and 2 list these and other parasite proteins that have been reported to be present at the host-exposed tegumental surface of intravascular life stages. It is by no means clear that these proteins are continually exposed at the surface once the parasites have invaded a vertebrate host; in most cases evidence for host exposure comes from experiments limited to one or a few life stages e.g. either schistosomula, lung worms or mature adult parasites and usually from a single schistosome species. The evidence for exposure of

Table 6-1. Schistosome Tegumental Surface Proteins

MOLECULE	EVIDENCE FOR SURFACE LOCATION	CHARACTERISTICS	REFS.
NUTRIENT TRANSPORTERS			
Glucose Transporter Protein (SGTP4)	Surface labeling, immunofluorescence, and immuno-EM	55 kDa facilitated diffusion sugar transporter	[6,8,9,29]
Permease Light Chain (SPRM11c)	Surface labeling and immunofluorescence	54 kDa amino acid transporter, heterodimer	[10,11]
ENZYMES			
Alkaline phosphatase	Surface enzyme activity, immunofluorescence and PIPLC release	Tetramer of 65 kDa subunits, GPI-linked	[12,13,15, 16,30]
Nucleoside-diphosphatase	Surface enzyme activity	Uncharacterized	[31]
ATP-diphosphohyrolase	Surface enzyme activity	63 kDa enzyme	[32]
Acetylcholinesterase	Surface enzyme activity, immunofluorescence and PIPLC release	Dimer of 110 kDa and 76 kDa subunits, GPI-linked	[14,18-20,33-36]
Esterase	Immunofluorescence	27 kDa non-specific esterase	[37]
m28	PIPLC release	28 kDa serine protease, GPI-linked	[38]
Leucine aminopeptidase	Surface enzyme activity	Surface enzyme uncharacterized	[39]
Triose phosphate isomerase	Immunofluorescence	28 kDa glycolytic enzyme	[40]
Glyceraldehyde-3-phosphate dehydrogenase	Immunofluorescence	37 kDa glycolytic enzyme	[41]

each molecule at the parasite surface comes from a number of sources, notably: an ability to label the molecule on intact parasites, an ability to visualize the molecule at the surface by immunofluorescence, immunohistochemistry or immuno-electron microscopy (EM), an ability to release the molecule from intact parasites using e.g. phosphatidylinositol phospholipase C (PIPLC) treatment or, for several enzymes, an ability to detect activity directly at the surface.

3.1   Nutrient Transporters

Schistosomes in the bloodstream transport glucose and other nutrients across their outer tegument [42]. This implies the presence of host-exposed **glucose transporter proteins** at the surface. A family of three different transporters from *S. mansoni,* designated schistosome glucose transporter protein (SGTP) 1, SGTP2 and SGTP4, have been cloned and characterized [6,8,9,29]. These proteins possess structural features common to the energy independent family of transporter proteins that facilitate glucose diffusion into cells. SGTP1, SGTP2 and SGTP4 are all predicted to be ~55 kDa, to possess 12 transmembrane domains, characteristic conserved amino acid motifs as well as a single potential glycosylation site between the first and second membrane-spanning domains. SGTP1 and SGTP4 are closely related, exhibiting 61% sequence identity [43]. Using the *Xenopus* expression system, it has been shown that SGTP1 and SGTP4 are functional facilitated diffusion glucose transporters; SGTP2 is not functional in this assay. The K_m for glucose uptake in *Xenopus* oocytes expressing schistosome GTPs is 1.3 mM for SGTP1 and 2 mM for SGTP4 [43]. Isolated adult parasites *in vitro* exhibit glucose transporting characteristics that are broadly similar to the biochemical traits of these cloned proteins [44].

SGTP4 localizes uniquely and specifically to the tegument of adult males and females. Using confocal microscopy and immuno-EM, the tegumental localization for SGTP1 and SGTP4 has been shown to be distinctly asymmetric. SGTP1 is found only in the basal membranes and SGTP4 only in the apical membranes and in both types of tegumental secretory bodies [8]. SGTP4, but not SGTP1, can be readily labeled by aqueous biotinylating agents on the surface of living worms, further demonstrating that it is present at the host/parasite interface. SGTP4 is detected by immuno-EM localization in both of the apical lipid bilayers, suggesting that this protein functions to import sugar from the glucose-rich host blood stream across both of the apical surface membranes and into the tegumental cytoplasm [8]. SGTP1 then transports some of this tegumental glucose across the basal membrane for use by the underlying tissues. No SGTP localizes to the intestinal tract, supporting the notion that sugar entry into the intravascular parasites occurs across the body surface and not through the gut.

SGTP4 is found only in intra-vertebrate stage parasites. Following host invasion, SGTP4 is detected first within the body of the parasite in a network of interconnected cytons [9]. As the parasite transforms into the intra-mammalian stage, SGTP4 is then detected in the cytoplasmic connections to the surface and next, it is deposited at the surface in discrete patches [9]. Over the following several hours, as the patches coalesce, the protein is finally deposited in a contiguous layer over the worm surface [9]. This swift movement of SGTP4 to the surface of the invading schistosomulum reflects the vital importance of prompt sugar import to replenish depleted reserves and to promote parasite survival.

Studies on amino acid import into adult male schistosomes have suggested the presence of at least five distinct amino acid uptake systems [45]. An **amino acid transporter** (or **permease**) has been identified in *S. mansoni* that likely represents the molecular basis of one of the amino acid uptake systems described in whole worms [10]. This protein, designated SPRM1lc (for Schistosome Permease 1 light chain), is a ~54 kDa member of a relatively new family of heterodimeric amino acid permeases [10]. These proteins associate with a type II glycoprotein (the heavy chain) to facilitate amino acid uptake [11]. SPRM1lc is capable of transporting several amino acids when it is expressed in *Xenopus* oocytes together with a heterologous heavy chain. The basic amino acids, histidine, arginine and lysine are all transported into oocytes via SPRM1lc. In addition, leucine, phenylalanine, methionine and glutamine are taken up [10]. The protein is detected in all schistosome life stages examined, suggesting that it acts as a housekeeping protein. In adult parasites, the protein is widely distributed in several tissues: SPRM1lc is detected throughout the parenchyma as well as in the tegument, the tegumental cytons and the connections between the two. Confocal microscopy reveals that SPRM1lc is located in the outer tegument where it likely functions to import amino acids across the surface of the tegument from the host bloodstream. The wide distribution of SPRM1lc within the adult worm suggests that the protein is also important in distributing amino acids throughout the organism.

Since the intra-mammalian stage parasites are known to transport other metabolites across their teguments, doubtless many more host-exposed nutrient transporters await discovery [45,46].

3.2 Surface Enzymes

A wide variety of proteins with enzymatic activity have been identified at the schistosome surface. Over 40 years ago, **alkaline phosphatase** activity was histochemically located in the tegument, as well as in several internal tissues, of adult schistosomes [12].

The *S. mansoni* alkaline phosphates have been purified from detergent extracts of adult parasites following ConA agarose affinity chromatography and gel filtration. It is a glycoprotein of ~260 kDa composed of four enzymatically inactive subunits of 65 kDa [15]. A monoclonal antibody (MAb) against the enzyme binds to the adult tegument [16]. Most alkaline phosphatase activity (~70%) is contained in tegument enriched fractions of both males and females [13]. Activity is detected in the material released from cultured schistosomula following treatment with the enzyme PIPLC. This suggests that some alkaline phosphatase is lipid anchored at the surface. The enzyme is exposed to the host and is immunogenic since most infected humans show a clear antibody response against it [30]. The role of tegumental alkaline phosphatase is uncertain, but has been suggested to be related to purine recovery and/or the regulation and transport of phosphate ions across the tegument [16]. Other phosphatase activities have been detected at the parasite surface and the relationships between them all has not yet been resolved [31,32].

More than 30 years ago, **esterase** activity was identified histochemically in various schistosome tissues including the adult tegument [17]. Some of the molecules responsible for this surface esterase activity have now been characterized. **Acetylcholinesterase** (AChE) activity in intravascular schistosomes is found not only in the neuromuscular system but also at the surface [14]. Anti-AChE antibodies stain the outer membranes [18,33]. Enzyme analysis suggests that about 50% of the total parasite AChE is present at the surface of intact schistosomula [18]. A substantial amount of AChE is glycosylphosphatidylinositol (GPI) anchored since it can be released from living parasites by PIPLC [19,34,35]. The surface AChE exists as a dimeric, globular form consisting of 110 kDa and 76 kDa subunits with additional, larger multimeric forms associated with internal tissues [36]. AChE is a natural antigen, since most infected experimental animals and humans show an antibody response against it [33].

A cDNA encoding an acetylcholinesterase from *S. haematobium* has been cloned and characterized [20]. The enzyme is predicted to be ~80 kDa. Sequence identities between it and homologs from *C. elegans, D. melanogaster* or humans are similar and vary from 29 to 33%. Polyclonal antibodies, raised against a synthetic peptide, confirm the localization of the enzyme at the worm surface and underlying muscle tissue [20]. A 27 kDa enzyme with non-specific esterolytic activity has also been described at the parasite surface [37]. The physiological function of either of these esterases at the schistosome surface is not known.

Table 6-2. Schistosome Tegumental Surface Proteins cont.,

MOLECULE	EVIDENCE FOR SURFACE LOCATION	CHARACTERISTICS	REFS.
RECEPTORS/RECEPTOR HOMOLOGS			
Receptor serine threonine kinase homolog (SmRK-1)	Surface labeling and immunofluorescence	66 kDa TGF receptor	[47,48]
Acetylcholine receptor	Receptor ligand, - bungarotoxin, binds to the surface	Uncharacterized	[49]
Low-density lipoprotein (LDL) receptors	Labeled LDL binds to the parasite surface	No consensus	[31,50-53]
Complement C2 receptor (CRIT)	Immuno-EM	32 kDa C2 binding protein	[21]
Complement C3 receptor	Conflicting reports of C3 or anti-C3 antisera binding parasite surface	130 kDa C3 binding protein	[22]
Complement C8, C9 receptor (SCIP1 / paramyosin)	Immunofluorescence, Immuno-EM and PIPLC release	97 kDa C8, C9 binding protein, GPI-linked	[23]
Fc receptor/ paramyosin	Conflicting labeling/ surface immunofluorescence localization. Conflicting reports of immunoglobulin binding at the surface	97 kDa muscle protein	[6,23,54-60]
STRUCTURAL PROTEINS/PROTEINS OF UNKNOWN ORIGIN			
Sm23	Immunofluorescence	23 kDa antigen, some fraction GPI-linked	[24,61,62]
Spine glycoprotein	Immunofluorescence	170 kDa glycoprotein	[63]
Sm200	Immunofluorescence and PIPLC release	200 kDa glycoprotein, GPI-linked	[27,28,64]

3.3   Receptors or Receptor Homologs

As the primary site of host/parasite contact, it is not surprising that schistosomes possess receptors at their surface that play a role in environmental sensing and host interaction. A number of surface receptors or receptor homologs have been described. A cDNA clone encoding a 594 amino acid schistosome **receptor serine threonine kinase homolog** from the transforming growth factor (TGF) super family has been identified [47]. The protein, designated **SmRK-1**, localizes to the tubercles found on the dorsal surface of male worms. SmRK1 exhibits 58% sequence identity with kinase domains of other receptors in the TGF family and contains a conserved glycine-serine motif [47]. It may function in signal transduction across the surface membrane once the parasites have reached the hepatic portal circulation and perhaps in male-female interactions. In support of a signaling role for SmRK1, human TGF binds and activates the receptor when it is expressed in a heterologous system [48]

The presence of a tegumental **receptor for low density lipoprotein** (LDL) at the schistosome surface has been suggested, since labeled LDL binds to parasites [50,51]. Purified human LDL binds to low molecular weight bands of ~16-18 kDa, on blots of schistosomula extracts and a 43 kDa protein in adult extracts [51]. Others find that human LDL binds to adult schistosome tegumental proteins of 14, 35 and 60 kDa [52]. A 207 amino acid, putative LDL-binding protein was identified in *S. japonicum* that immunolocalizes to the adult tegument [53]. The contributions of each of these numerous LDL receptor candidates to lipid binding by the parasites and whether all are exposed at the parasite surface have yet to be clarified. The physiological significance of surface LDL receptors is unclear since LDL, or other material, bound at the schistosome surface has never been shown to be internalized through endocytosis.

Schistosomes have been reported to possess several **complement receptors** at their surface that are reported to be capable of binding to various components of the host's complement cascade [21-23]. In this way, the parasites are proposed to inhibit the formation of membrane attack complexes that could damage or destroy the tegumental membranes. Evidence for the presence of a C2-binding protein [21], a C3 binding protein [22] and a C8 and C9 binding protein [23] have all been reported at the surface of intravascular schistosomes.

Because of the reported ability of adult schistosomes to adsorb heterospecific immunoglobulin onto their tegumental surfaces via their Fc domains (see "Immunoglobulin" later), efforts were made to identify this **Fc receptor**. First, freshly perfused adult parasites were surface biotinylated and extracts were then incubated with human IgG-Fc coupled to sepharose. A 97 kDa biotinylated protein bound to the sepharose complex and was identified as paramyosin [60]. Paramyosin-IgG complexes were detected on the surface of fresh adult *S. mansoni*, using protein G-sepharose [60]. The localization of paramyosin outside of schistosome muscle tissue is controversial. While some workers report that the protein is present in the tegument of adult parasites [23,54-56], others dispute this [6,57,58]. Work with *S. japonicum* indicates that paramyosin localizes to the surface of lung stage worms but not to the surface or the tegument of adult parasites [59].

3.4   Structural Proteins Or Those Of Unknown Function

A number of proteins have been identified at the parasite surface that act as structural proteins or have currently no known function. These include the 218 amino acid membrane protein, **Sm23**. This protein was originally identified as the target of a MAb that had been prepared using mice immunized with a surface membrane enriched schistosomula preparation [24]. Sm23 is widely distributed in all developmental stages and, in intravascular stage parasites, is found in several tissues including the parasite surface [24]. Some forms of Sm23 contain a GPI anchor [61]. The protein is immunogenic in infected humans and rabbits [62]. Immunization of mice with Sm23 has been shown to induce a protective immune response in several vaccine trials [65,66]. Immunization of experimental animals with the *S. japonicum* homolog, Sj23, evokes protective immunity in some, but not all, instances [67,68]. Sm23 exhibits strong "domain homology" with proteins of the "tetraspannin" family, suggesting that the protein may be involved in signal transduction across the schistosome surface [69].

Mice that have been exposed to radiation-attenuated cercariae are highly resistant to challenge infection. Among the antigens recognized by sera from these protected mice is a 200 kDa glycoprotein, **Sm200** [27]. Mice vaccinated with a recombinant 62 kDa fragment of this molecule (designated IrV-5) exhibit variable levels of protection from challenge infection [27]. Sera from these mice, as well as a mAb generated against purified recombinant IrV-5, bind to the surface of schistosomula [27]. The 200 kDa protein is available for biotinylation on intact schistosomula and is immunoprecipitated using these sera [27]. In addition, the 200 kDa protein can be released from the adult surface following PIPLC treatment [64] and a GPI anchoring signal is identified in the predicted sequence of the 200 kDa protein [28]. Over its whole length, the protein exhibits no significant homology with other database sequences and the function this protein plays at the parasite surface is unknown [28].

3.5   Surface Tegumental Lipid Composition

Adult schistosome tegumental membrane extracts contain high concentrations of phospholipids and cholesterol [70]. The phospholipid composition of surface membrane extracts from 6 week adult parasites shows a typical plasma membrane profile with a high sphingomyelin content ~20%, [71]. Younger forms (2-3 weeks old) have negligible amounts of sphingomyelin [70]. Phosphatidylcholine and phosphatidyl ethanolamine are major phospholipid constituents of the surface membranes [72].

3.6   Surface Tegumental Carbohydrate Composition

The abundance of carbohydrate at the schistosome surface has been demonstrated both by lectins binding to parasites and by detection of large numbers of glycoproteins in tegumental extracts [73,74]. In seminal work, Schmidt reported intense binding of a number of gold-labeled lectins to the adult schistosome surface [75]. Both male and female parasites are entirely and evenly covered with glycans that appear as an amorphous mucus of low electron density. The precise biochemical nature of the glycans that compose this covering is unclear; lectin-binding studies indicate large amounts of exposed N-acetylactosamine (i.e. galactose (1-4) N- acetylglucosamine) [75]. These parasite surface molecules, being common components of vertebrate glycans, may not be immunogenic and may therefore mask the parasites against immune attack [75].

The fucose-containing trisaccharide Gal   1-4(Fuc 1-3)GlcNAc 1 (designated Lewis^x (Le^x) or CD15) has been located very widely at the surface of intra-mammalian stage schistosomes using anti- Le^x MAbs [76,77]. Le^x is also detected in the adult gut and on material released by the parasites [76,77]. Le^x elicits an immune response in infected hosts [76-78]. A MAb antibody against Le^x directs effector eosinophils to kill schistosomula *in vitro* [79]. These data show that Le^x is available for antibody binding and is exposed at the parasite surface.

4.   HOST MOLECULES

The presence of epitopes shared between parasite and host was first suggested by the ability of anti-schistosome antiserum raised in rabbits to bind to host mouse serum molecules [80]. Host molecules on the parasite surface were later demonstrated by innovative worm transfer experiments. When adult worms are transferred from a mouse to a monkey that has been immunized against mouse red blood cells (an "anti-mouse" monkey), most of the parasites do not survive the transfer. In contrast, adult worms transferred from a mouse to control, non-immunized monkeys (or to monkeys immunized against sheep red blood cells) survive well [81,82]. Transferred parasites have severely damaged teguments, with vacuolated or destroyed syncytia [81], indicating that most or all of the acquired host antigens are located at the tegumental surface. Immunization with mouse erythrocytes or mouse erythrocyte ghosts provides greatest protection (90-100%) but immunization with mouse erythrocyte soluble proteins also provides considerable protection (74-100%) [82]. These transfer experiments were carried out on relatively small numbers of experimental animals

[81,82]. Other workers report that protection is not consistently seen in similar experiments involving other host species or different mouse strains [83,84]. Nonetheless, such transfer experiments established the notion that host cell molecules could become attached to the parasite and thereby provide a target for immunological killing in the immunized animals.

That freshly perfused adult worms have host molecules associated with them was determined using rabbit anti-whole-mouse-extract antisera in a mixed agglutinin test [85] or by immunofluorescence [86]. While the precise molecules that are the targets of immunological attack at the adult parasite surface have not been elucidated, a variety of host molecules have been demonstrated, or implicated, to be present there. They belong to several classes and include blood group antigens [87, 88], MHC molecules [89-91] and immunoglobulin [92-98]. Table 2 summarizes what is known about these and other host proteins that have been reported to be present at the tegumental surface of intravascular stage schistosomes.

Table 6-3. Surface Proteins of Host Origin

MOLECULE	EVIDENCE FOR SURFACE LOCATION	COMMENTS	REFS.
Blood Group Antigens	Agglutination, immunofluorescence and worm transfer experiments	A, B and Lewis[b] antigens detected; M, N and Duffy antigens not detected.	[87,88,99]
MHC molecules	Immunofluorescence	MHC class I and II alloantigens detected	[89-91]
Immunoglobulin	Conflicting reports of surface immunoglobulin binding	Some studies detect IgA, IgM, IgG1, IgG2a, IgG2b, IgG3 isotypes	[89,93-95,97,98]
Delay Accelerating Factor	Surface labeling and immunofluorescence	70 kDa molecule, inhibits complement	[100,101]
Skin antigens	Immunofluorescence	Uncharacterized	[102]
Fibronectin	Immunofluorescence/immuno histochemistry	Whether the molecules are of parasite or host origin is unresolved	[103]
•2-macroglobulin	Agglutination and immuno-EM	Whether the molecules are of parasite or host origin is unresolved	[104]

4.1 Blood Group Antigens

Schistosomula, cultured with erythrocytes or erythrocyte extracts, can adsorb blood group A and B antigens onto their surfaces, as determined by means of a mixed agglutination test [87]. Since formalin-fixed, dead schistosomula also acquire antigen, it seems that these substances can be passively adsorbed. No uptake of several other blood group antigens H, M, N, P, C, D, E, P or Duffy, could be demonstrated [87].

In other experiments, schistosomula were cultured with erythrocytes of the AB Rh- blood group. Parasites were then surgically transferred into the vasculature of rhesus monkeys that had been immunized with the same erythrocytes. Most parasites were destroyed following this procedure (95-98%) whereas controls were not, suggesting that the acquired host antigens formed targets for an immune response [99]. In later work, schistosomula were cultured in medium containing blood from A secretor or B secretor humans [88]. The schistosomula were transferred by surgery into the vascular system of monkeys immunized with blood type A glycoproteins. Those schistosomula that had been cultured in type A blood were almost totally destroyed following transfer, while those cultured in B type blood survived well. Again, it was inferred that the A antigens provided a focus for immune-mediated killing *in vivo* [88]. Mixed agglutination and immunofluorescence analysis showed that, in addition to A and B antigens, H and Lewis[b+] antigens could be acquired by parasites during culture, whereas Rhesus, M, N, S and Duffy antigen could not [88].

Whether schistosomes adsorb host blood group antigen *in vivo* is unresolved. If they do, these antigens would be good candidates as targets of schistosome killing in adult worm transfer experiments [81,82].

4.2   Major Histocompatibility (MHC) Molecules

S. mansoni schistosomula, recovered from the lungs of inbred mice, possess serologically detectable alloantigens on their tegumental surfaces [89]. Using congenic mice and appropriate antisera, gene products of the K and I sub-regions of the major histocompatability complex were demonstrated among those alloantigens acquired by the parasite. The K and I regions encode MHC class I and MHC class II glycoproteins. The acquisition of these molecules is selective since parasites do not stain for other known cell surface antigens including Thy 1, Ly 1 and H-Y [89].

MAbs recognizing H-2K[k] bind to the surface of lung-stage schistosomes [90,91]. Furthermore, these antibodies precipitate a ~45 kDa molecule from surface-labeled lung-stage parasites and labeled spleen cells [91]. Using a variety of anti-alloantigen antisera, gene products from other regions of the MHC region have been demonstrated on the surface of lung stage parasites [90]. These include products from the H-2D, H-2K, I-E and I-A regions [90].

Schistosomula recovered from the lungs of mice and reinjected into allogeneic recipients can exchange their alloantigens. This exchange process begins by 15 hours after reinjection. By 87 hours after reinjection, all of the parasites express surface alloantigens of their new hosts [89].

4.3   Immunoglobulin

The ability of adult parasites to adsorb heterospecific antibody onto their tegumental surfaces was shown by several groups, though occasionally with conflicting results [89,93,94,97]. In early work, mice were first immunized against red blood cells (rbcs) or bovine serum albumin or horseradish peroxidase. Next, worms recovered from those animals and incubated in their respective antigens selectively bound to their tegumental surfaces only those antigens to which their murine host had been immunized [93]. In related work, schistosomula were incubated with mouse anti-sheep rbc antibody and then with sheep rbcs. In a process called rosetting, the rbcs coat the parasites. Here antibody binding to an Fc receptor (present at the parasite surface) exposes antigen-binding sites that are available to adhere to the rbcs [94]. Five day old schistosomula and adult worms no longer express the ability to form rosettes, suggesting (in contrast to the work noted above [93]) the absence of exposed Fc receptors in these more mature parasites [94].

IgG1, IgG2a, and IgG2b are all detected with a general, homogeneous distribution on the outer tegumental membrane of adult parasites, using [95]. These antibodies are also detected in worm eluates [95] and several antibody classes (IgG, IgG1, IgG2a, IgG2b, IgG3, IgM, IgA) are detected in material released from cultured adult worms recovered from infected mice [98].

Some workers readily detect bound immunoglobulin at the adult parasite surface [93,95] but others do not [89,94]. This is in keeping with the ability of some researchers to detect surface Fc receptor (paramyosin) [23,60] and the inability of others to so do [6,58]; suggesting that receptor accessibility and thus the ability of antibody to bind to the surface, may vary depending on parasite strain, parasite age or culture and experimental conditions.

Table 2 lists several host molecules, in addition to those just described, that have been reported to associate with the parasite surface. It has been suggested that schistosome acquisition of such host molecules permits the parasite to mask its own antigens and thereby help prevent its recognition by the host [105]. Host IgG for instance, bound to the parasite surface via the Fc region, has been proposed to sterically hinder parasite-specific antibody from binding to appropriate antigens [97]. Currently however, there is no direct evidence to support the notion that host molecules mask parasite antigens to provide protection and it may be that the acquisition of host molecules has no functional significance for the parasite but is merely a consequence of the chemical and biophysical nature of its outer surface.

5.   TEGUMENTAL SUBSTRUCTURE

It is important to note that the host-interactive schistosome surface is not a molecularly uniform structure. By tracking the movements of fluorescent lipid analogues that have been applied to the adult schistosome surface, it has been concluded that the surface membrane is differentiated into domains of varying composition and structure [106,107]. The diffusion rates of different lipids in the outer membrane can vary substantially suggesting that lipids may be sequestered into specialized micro-domains in the outer tegumental membranes with diverse properties [108]. Different membrane domains may have specific functions; specialized invaginations of the plasma membrane forming caveolae-like structures have been described at the adult schistosome surface [26]. By analogy with the role of caveolae in mammalian cells, the caveolae-like structures described in the schistosome surface membrane may serve as specialized foci for signaling or the uptake of metabolites [26]. How the different parasite and host molecules described in this review are arranged relative to one another in the membranous microdomains of the host-interactive surface remains to be determined.

ACKNOWLEDGEMENTS

Thanks to Dr. Akram Da'dara for helpful discussions and critical review of this manuscript and to Dr. D. Harn for his support of the work. This work was supported by National Institutes of Health Grant R21 AI47453.

REFERENCES

1. McLaren, D.J. and Hockley, D.J., 1977, Blood flukes have a double outer membrane, *Nature* **269**:147-149.
2. Threadgold L.T., 1984, "Parasitic platyhelminths". In *Biology of the integument*, J Bereiter-Hahn, AG Matolsky and KS Richards, eds. Berlin, Springer-Verlag, pp. 132-191.
3. Wilson, R.A. and Barnes, P.E., 1974, The tegument of *Schistosoma mansoni*: observations on the formation, structure and composition of cytoplasmic inclusions in relation to tegument function, *Parasitology* **68**:239-258.
4. Smith J.H., Reynolds, E.S. and Von Lichtenberg, F., 1969, The integument of *Schistosoma mansoni*, *Am J Trop Med Hyg* **18**:28-49.
5. Hockley D.J. and McLaren, D.J., 1973, *Schistosoma mansoni*: changes in the outer membrane of the tegument during development from cercaria to adult worm, *Int J Parasitol* **3**:13-25.

6. Skelly P.J. and Shoemaker, C.B., 1996, Rapid appearance and asymmetric distribution of glucose transporter SGTP4 at the apical surface of intramammalian-stage *Schistosoma mansoni*, *Proc Natl Acad Sci U S A* **93**:3642-3646.

7. Verjovski-Almeida S., Demarco, R., Martins, E.A., *et al.*, 2003, Transcriptome analysis of the acoelomate human parasite *Schistosoma mansoni*, *Nat Genet* **35**:148-157.

8. Jiang J., Skelly, P.J., Shoemaker, C.B., *et al.*, 1996, *Schistosoma mansoni*: the glucose transport protein SGTP4 is present in tegumental multilamellar bodies, discoid bodies, and the surface lipid bilayers, *Exp Parasitol* **82**:201-210.

9. Skelly P.J. and Shoemaker, C.B., 2001, The *Schistosoma mansoni* host-interactive tegument forms from vesicle eruptions of a cyton network, *Parasitology* **122**: 67-73.

10. Skelly P.J., Pfeiffer, R., Verrey, F., *et al.*, 1999, SPRM1lc, a heterodimeric amino acid permease light chain of the human parasitic platyhelminth, *Schistosoma mansoni*, *Parasitology* **119**: 569-576.

11. Mastroberardino L., Spindler, B., Pfeiffer, R., *et al.*, 1998, Amino-acid transport by heterodimers of 4F2hc/CD98 and members of a permease family, *Nature* **395**:288-291.

12. Dusanic D.G., 1959, Histochemical observations of alkaline phosphatase in *Schistosoma mansoni*, *J Infect Dis* **105**:1-8.

13. Cesari I.M., 1974, *Schistosoma mansoni*: distribution and characteristics of alkaline and acid phosphatase, *Exp Parasitol* **36**:405-414.

14. Levi-Schaffer F., Tarrab-Hazdai, R., Schryer, M.D., *et al.*, 1984, Isolation and partial characterization of the tegumental outer membrane of schistosomula of *Schistosoma mansoni*, *Mol Biochem Parasitol* **13**:283-300.

15. Payares G., Smithers, S.R. and Evans, W.H., 1984, Purification and topographical location of tegumental alkaline phosphatase from adult *Schistosoma mansoni*, *Mol Biochem Parasitol* **13**:343-360.

16. Pujol F.H., Liprandi, F., Rodriguez, M., *et al.*, 1990, Production of a mouse monoclonal antibody against the alkaline phosphatase of adult *Schistosoma mansoni*, *Mol Biochem Parasitol* **40**:43-52.

17. Fripp P.J., 1967, Histochemical localization of esterase activity in Schistosomes, *Exp Parasitol* **21**:380-390.

18. Espinoza B., Parizade, M., Ortega, E., *et al.*, 1995, MAbs against acetylcholinesterase of *Schistosoma mansoni*: production and characterization, *Hybridoma* **14**:577-586.

19. Camacho M., Alsford, S. and Agnew, A., 1996, Molecular forms of tegumental and muscle acetylcholinesterases of Schistosoma, *Parasitology* **112**:199-204.

20. Jones A.K., Bentley, G.N., Oliveros Parra, W.G., *et al.*, 2002, Molecular characterization of an acetylcholinesterase implicated in the regulation of glucose scavenging by the parasite Schistosoma, *Faseb J* **16**:441-443.

21. Inal J.M., Schneider, B., Armanini, M., *et al.*, 2003, A peptide derived from the parasite receptor, complement C2 receptor inhibitor trispanning, suppresses immune complex-mediated inflammation in mice, *J Immunol* **170**:4310-4317.

22. Silva E.E., Clarke, M.W. and Podesta, R.B., 1993, Characterization of a C3 receptor on the envelope of *Schistosoma mansoni*, *J Immunol* **151**:7057-7066.

23. Deng J., Gold, D., Loverde, P.T., *et al.*, 2003, Inhibition of the complement membrane attack complex by *Schistosoma mansoni* paramyosin, *Infect Immun* **71**:6402-6410.

24. Harn D.A., Mitsuyama, M., Huguenel, E.D., *et al.*, 1985, *Schistosoma mansoni*: detection by monoclonal antibody of a 22,000-dalton surface membrane antigen which may be blocked by host molecules on lung stage parasites, *J Immunol* **135**:2115-2120.

25. Wright M.D., Henkle, K.J. and Mitchell, G.F., 1990, An immunogenic Mr 23,000 integral membrane protein of *Schistosoma mansoni* worms that closely resembles a human tumor-associated antigen, *J Immunol* **144**:3195-3200.

26. Racoosin E.L., Davies, S.J. and Pearce, E.J., 1999, Caveolae-like structures in the surface membrane of *Schistosoma mansoni*, *Mol Biochem Parasitol* **104**:285-297.

27. Soisson L.M., Masterson, C.P., Tom, T.D., *et al.*, 1992, Induction of protective immunity in mice using a 62-kDa recombinant fragment of a *Schistosoma mansoni* surface antigen, *J Immunol* **149**:3612-3620.

28. Hall T.M., Joseph, G.T. and Strand, M., 1995, *Schistosoma mansoni*: molecular cloning and sequencing of the 200-kDa chemotherapeutic target antigen, *Exp Parasitol* **80**:242-249.

29. Skelly P.J. and Shoemaker, C.B., 2000, Induction cues for tegument formation during the transformation of *Schistosoma mansoni* cercariae, *Int J Parasitol* **30**:625-631.

30. Pujol F.H. and Cesari, I.M., 1990, Antigenicity of adult *Schistosoma mansoni* alkaline phosphatase, *Parasite Immunol* **12**:189-198.

31. Bogitsh B.J.and Krupa, P.L., 1971, *Schistosoma mansoni* and Haematoloechus medioplexus: nuclosidediphosphatase localization in tegument, *Exp Parasitol* **30**:418-425.

32. Vasconcelos E.G., Nascimento, P.S., Meirelles, M.N., *et al.*, 1993, Characterization and localization of an ATP-diphosphohydrolase on the external surface of the tegument of *Schistosoma mansoni*, *Mol Biochem Parasitol* **58**:205-214.

33. Espinoza B., Tarrab-Hazdai, R., Himmeloch, S., *et al.*, 1991, Acetylcholinesterase from *Schistosoma mansoni*: immunological characterization, *Immunol Lett* **28**:167-174.

34. Espinoza B., Tarrab-Hazdai, R., Silman, I., *et al.*, 1988, Acetylcholinesterase in *Schistosoma mansoni* is anchored to the membrane via covalently attached phosphatidylinositol, *Mol Biochem Parasitol* **29**:171-179.

35. Espinoza B., Silman, I., Arnon, R., *et al.*, 1991, Phosphatidylinositol-specific phospholipase C induces biosynthesis of acetylcholinesterase via diacylglycerol in *Schistosoma mansoni*, *Eur J Biochem* **195**:863-870.

36. Camacho M., Tarrab-Hazdai, R., Espinoza, B., *et al.*, 1994, The amount of acetylcholinesterase on the parasite surface reflects the differential sensitivity of schistosome species to metrifonate, *Parasitology* **108**:153-160.

37. Doenhoff M.J., Modha, J. and Lambertucci, J.R., 1988, Anti-schistosome chemotherapy enhanced by antibodies specific for a parasite esterase, *Immunology* **65**:507-510.

38. Ghendler Y., Parizade, M., Arnon, R., *et al.*, 1996, *Schistosoma mansoni*: evidence for a 28-kDa membrane-anchored protease on schistosomula, *Exp Parasitol* **83**:73-82.

39. Fripp P.J., 1967, The histochemical localization of leucine aminopeptidase activity in Schistosoma rodhaini, *Comp Biochem Physiol B* **20**:307-309.

40. Harn D.A., Gu, W., Oligino, L.D., *et al.*, 1992, A protective monoclonal antibody specifically recognizes and alters the catalytic activity of schistosome triose-phosphate isomerase, *J Immunol* **148**:562-567.

41. Goudot-Crozel V., Caillol, D., Djabali, M., *et al.*, 1989, The major parasite surface antigen associated with human resistance to schistosomiasis is a 37-kD glyceraldehyde-3P-dehydrogenase, *J Exp Med* **170**:2065-2080.

42. Skelly P.J., Tielens, A.G.M. and Shoemaker, C.B., 1998, Glucose transport and metabolism in mammalian-stage schistosomes, *Parasitology Today* **14**:402-406.

43. Skelly P.J., Kim, J.W., Cunningham, J., *et al.*, 1994, Cloning, characterization, and functional expression of cDNAs encoding glucose transporter proteins from the human parasite *Schistosoma mansoni*, *J Biol Chem* **269**:4247-4253.

44. Uglem G.L.and Read, C.P., 1975, Sugar transport and metabolism in *Schistosoma mansoni*, *J Parasitol* **61**:390-397.

45. Asch H.L. and Read, C.P., 1975, Membrane transport in *Schistosoma mansoni*: transport of amino acids by adult males, *Exp Parasitol* **38**:123-135.

46. Levy M.G.and Read, C.P., 1975, Purine and pyrimidine transport in *Schistosoma mansoni*, *J Parasitol* **61**:627-632.

47. Davies S.J., Shoemaker, C.B. and Pearce, E.J., 1998, A divergent member of the transforming growth factor beta receptor family from *Schistosoma mansoni* is expressed on the parasite surface membrane, *J Biol Chem* **273**:11234-11240.

48. Beall M.J. and Pearce, E.J., 2001, Human transforming growth factor-beta activates a receptor serine/threonine kinase from the intravascular parasite *Schistosoma mansoni*, *J Biol Chem* **276**:31613-31619.

49. Camacho M., Alsford, S., Jones, A., *et al.*, 1995, Nicotinic acetylcholine receptors on the surface of the blood fluke Schistosoma, *Mol Biochem Parasitol* **71**:127-134.

50. Rumjanek F.D., Campos, E.G. and Afonso, L.C., 1988, Evidence for the occurrence of LDL receptors in extracts of schistosomula of *Schistosoma mansoni*, *Mol Biochem Parasitol* **28**: 145-152.

51. Xu X. and Caulfield, J.P., 1992, Characterization of human low density lipoprotein binding proteins on the surface of schistosomula of *Schistosoma mansoni*, *Eur J Cell Biol* **57**:229-235.

52. Tempone A.J., Bianconi, M.L. and Rumjanek, F.D., 1997, The interaction of human LDL with the tegument of adult *Schistosoma mansoni*, *Mol Cell Biochem* **177**:139-144.

53. Fan J., Gan, X., Yang, W., *et al.*, 2003, A Schistosoma japonicum very low-density lipoprotein-binding protein, *Int J Biochem Cell Biol* **35**:1436-1451.

54. Pearce E.J., James, S.L., Dalton, J., *et al.*, 1986, Immunochemical characterization and purification of Sm-97, a *Schistosoma mansoni* antigen monospecifically recognized by antibodies from mice protectively immunized with a nonliving vaccine, *J Immunol* **137**:3593-3600.

55. Laclette J.P., Skelly, P.J., Merchant, M.T., *et al.*, 1995, Aldehyde fixation dramatically alters the immunolocalization pattern of paramyosin in platyhelminth parasites, *Exp Parasitol* **81**:140-143.

56. Matsumoto Y., Perry, G., Levine, R.J., *et al.*, 1988, Paramyosin and actin in schistosomal teguments, *Nature* **333**:76-78.

57. Schmidt J., Bodor, O., Gohr, L., *et al.*, 1996, Paramyosin isoforms of *S. mansoni* are phosphorylated and localized in a large variety of muscle types, *Parasitology* **112**:459-467.

58. Davies S.J. and Pearce, E.J., 1995, Surface-associated serine-threonine kinase in *Schistosoma mansoni*, *Mol Biochem Parasitol* **70**:33-44.

59. Gobert G.N., Stenzel, D.J., Jones, M.K., *et al.*, 1997, Schistosoma japonicum: immunolocalization of paramyosin during development, *Parasitology* **114**:45-52.

60. Loukas A., Jones, M.K., King, L.T., *et al.*, 2001, Receptor for Fc on the surfaces of schistosomes, *Infect Immun* **69**:3646-3651.

61. Koster B. and Strand, M., 1994, *Schistosoma mansoni*: Sm23 is a transmembrane protein that also contains a glycosylphosphatidylinositol anchor, *Arch Biochem Biophys* **310**:108-117.

62. Koster B., Hall, M.R. and Strand, M., 1993, *Schistosoma mansoni*: immunoreactivity of human sera with the surface antigen Sm23, *Exp Parasitol* **77**:282-294.

63. Norden A.P., Aronstein, W.S. and Strand, M., 1982, *Schistosoma mansoni*: identification, characterization, and purification of the spine glycoprotein by monoclonal antibody, *Exp Parasitol* **54**:432-442.

64. Sauma S.Y., Tanaka, T.M. and Strand, M., 1991, Selective release of a glycosylphosphatidylinositol-anchored antigen from the surface of *Schistosoma mansoni*, *Mol Biochem Parasitol* **46**:73-80.

65. Da'dara A.A., Skelly, P.J., Fatakdawala, M., *et al.*, 2002, Comparative efficacy of the *Schistosoma mansoni* nucleic acid vaccine, Sm23, following microseeding or gene gun delivery, *Parasite Immunol* **24**:179-187.

66. Da'dara A.A., Skelly, P.J., Walker, C.M., *et al.*, 2003, A DNA-prime/protein-boost vaccination regimen enhances Th2 immune responses but not protection following *Schistosoma mansoni* infection, *Parasite Immunol* **25**:429-437.

67. Waine G.J., Alarcon, J.B., Qiu, C., *et al.*, 1999, Genetic immunization of mice with DNA encoding the 23 kDa transmembrane surface protein of Schistosoma japonicum (Sj23) induces antigen-specific immunoglobulin G antibodies, *Parasite Immunol* **21**:377-381.

68. Shi F., Zhang, Y., Ye, P., *et al.*, 2001, Laboratory and field evaluation of Schistosoma japonicum DNA vaccines in sheep and water buffalo in China, *Vaccine* **20**:462-467.

69. Gaugitsch H.W., Hofer, E., Huber, N.E., *et al.*, 1991, A new superfamily of lymphoid and melanoma cell proteins with extensive homology to *Schistosoma mansoni* antigen Sm23, *Eur J Immunol* **21**:377-383.

70. Rogers M.V. and McLaren, D.J., 1987, Analysis of total and surface membrane lipids of *Schistosoma mansoni*, *Mol Biochem Parasitol* **22**:273-288.

71. Allan D., Payares, G. and Evans, W.H., 1987, The phospholipid and fatty acid composition of *Schistosoma mansoni* and of its purified tegumental membranes, *Mol Biochem Parasitol* **23**:123-128.

72. Roberts S.M., Aitken, R., Vojvodic, M., *et al.*, 1983, Identification of exposed components on the surface of adult *Schistosoma mansoni* by lactoperoxidase-catalysed iodination, *Mol Biochem Parasitol* **9**:129-143.

73. Torpier G. and Capron, A., 1980, Intramembrane particle movements associated with binding of lectins on *Schistosoma mansoni* surface, *J Ultrastruct Res* **72**:325-335.

74. Simpson A.J., Correa-Oliveira, R., Smithers, S.R., *et al.*, 1983, The exposed carbohydrates of schistosomula of *Schistosoma mansoni* and their modification during maturation in vivo, *Mol Biochem Parasitol* **8**:191-205.

75. Schmidt J., 1995, Glycans with N-acetyllactosamine type 2-like residues covering adult *Schistosoma mansoni*, and glycomimesis as a putative mechanism of immune evasion, *Parasitology* **111**:325-336.

76. Koster B. and Strand, M., 1994, *Schistosoma mansoni*: immunolocalization of two different fucose-containing carbohydrate epitopes, *Parasitology* **108**:433-446.

77. Remoortere A., Hokke, C.H., Van Dam, G.J., *et al.*, 2000, Various stages of schistosoma express Lewis(x), LacdiNAc, GalNAcbeta1-4 (Fucalpha1-3)GlcNAc and GalNAcbeta1-4(Fucalpha1-2Fucalpha1-3)GlcNAc carbohydrate epitopes: detection with monoclonal antibodies that are characterized by enzymatically synthesized neoglycoproteins, *Glycobiology* **10**: 601-609.

78. Nyame A.K., Pilcher, J.B., Tsang, V.C., *et al.*, 1996, *Schistosoma mansoni* infection in humans and primates induces cytolytic antibodies to surface Le(x) determinants on myeloid cells, *Exp Parasitol* **82**:191-200.

79. Ko A.I., Drager, U.C. and Harn, D.A., 1990, A *Schistosoma mansoni* epitope recognized by a protective monoclonal antibody is identical to the stage-specific embryonic antigen 1, *Proc Natl Acad Sci U S A* **87**:4159-4163.

80. Damian R.T., 1967, Common antigens between adult *Schistosoma mansoni* and the laboratory mouse, *J Parasitol* **53**:60-64.

81. Smithers S.R., Terry, R.J. and Hockley, D.J., 1969, Host antigens in schistosomiasis, *Proc R Soc Lond B Biol Sci* **171**:483-494.

82. Clegg J.A., Smithers, S.R. and Terry, R.J., 1970, "Host" antigens associated with schistosomes: observations on their attachment and their nature, *Parasitology* **61**:87-94.

83. Boyer M.H. and Ketchum, D.G., 1976, The host antigen phenomenon in experimental murine schistosomiasis. II. Failure to demonstrate destruction of parasites transferred from hamsters to mice, *J Immunol* **116**:1093-1095.

84. Boyer M.H., Ketchum, D.G. and Palmer, P.D., 1976, The host antigen phenomenon in experimental murine schistosomiasis: the transfer of 3-week old *Schistosoma mansoni* between two inbred strains of mice, *Int J Parasitol* **6**:235-238.

85. Sell K.W. and Dean, D.A., 1972, Surface antigens on *Schistosoma mansoni*. I. Demonstration of host antigens on schistosomula and adult worms using the mixed antiglobulin test, *Clin Exp Immunol* **12**:315-324.

86. Goldring O.L., Sher, A., Smithers, S.R., *et al.*, 1977, Host antigens and parasite antigens of murine *Schistosoma mansoni*, *Trans R Soc Trop Med Hyg* **71**:144-148.

87. Dean D.A., 1974, *Schistosoma mansoni*: adsorption of human blood group A and B antigens by schistosomula, *J Parasitol* **60**:260-263.

88. Goldring O.L., Clegg, J.A., Smithers, S.R., *et al.*, 1976, Acquisition of human blood group antigens by *Schistosoma mansoni*, *Clin Exp Immunol* **26**:181-187.

89. Sher A., Hall, B.F. and Vadas, M.A., 1978, Acquisition of murine major histocompatibility complex gene products by schistosomula of *Schistosoma mansoni*, *J Exp Med* **148**:46-57.

90. Gitter B.D. and Damian, R.T., 1982, Murine alloantigen acquisition by schistosomula of *Schistosoma mansoni*: further evidence for the presence of K, D, and I region gene products on the tegumental surface, *Parasite Immunol* **4**:383-393.

91. Simpson A.J., Singer, D., Mccutchan, T.F., *et al.*, 1983, Evidence that schistosome MHC antigens are not synthesized by the parasite but are acquired from the host as intact glycoproteins, *J Immunol* **131**:962-965.

92. Kemp W.M., Damian, R.T. and Greene, N.D., 1976, Immunocytochemical localization of IgG on adult *Schistosoma mansoni* tegumental surfaces, *J Parasitol* **62**:830-832.

93. Kemp W.M., Merritt, S.C., Bogucki, M.S., *et al.*, 1977, Evidence for adsorption of heterospecific host immunoglobulin on the tegument of *Schistosoma mansoni*, *J Immunol* **119**:1849-1854.

94. Torpier G., Capron, A. and Ouaissi, M.A., 1979, Receptor for IgG(Fc) and human beta2-microglobulin on *S. mansoni* schistosomula, *Nature* **278**:447-449.

95. Kemp W.M., Merritt, S.C. and Rosier, J.G., 1978, *Schistosoma mansoni*: identification of immunoglobulins associated with the tegument of adult parasites from mice, *Exp Parasitol* **45**:81-87.

96. Kemp W.M., Brown, P.R., Merritt, S.C., *et al.*, 1980, Tegument-associated antigen modulation by adult male *Schistosoma mansoni*, *J Immunol* **124**:806-811.

97. Tarleton R.L. and Kemp, W.M., 1981, Demonstration of IgG-Fc and C3 receptors on adult *Schistosoma mansoni*, *J Immunol* **126**:379-384.

98. Gearner G.W. and Kemp, W.M., 1994, Electrophoretic and serological analysis of host antigens associated with the adult *Schistosoma mansoni* tegument, *J Parasitol* **80**:275-283.

99. Clegg J.A., Smithers, S.R. and Terry, R.J., 1971, Acquisition of human antigens by *Schistosoma mansoni* during cultivation in vitro, *Nature* **232**:653-654.

100. Pearce E.J., Hall, B.F. and Sher, A., 1990, Host-specific evasion of the alternative complement pathway by schistosomes correlates with the presence of a phospholipase C-sensitive surface molecule resembling human decay accelerating factor, *J Immunol* **144**:2751-2756.

101. Horta M.F., Ramalho-Pinto, F.J. and Fatima, M., 1991, Role of human decay-accelerating factor in the evasion of *Schistosoma mansoni* from the complement-mediated killing in vitro, *J Exp Med* **174**:1399-1406.

102. Smith H.V. and Kusel, J.R., 1979, The acquisition of antigens in the intercellular substance of mouse skin by schistosomula of *Schistosoma mansoni*, *Clin Exp Immunol* **36**:430-435.

103. Ouaissi M.A., Cornette, J. and Capron, A., 1984, Occurrence of fibronectin antigenic determinants on *Schistosoma mansoni* lung schistosomula and adult worms, *Parasitology* **88**:85-96.

104. Kemp W.M., Damian, R.T., Greene, N.D., *et al.*, 1976, Immunocytochemical localization of mouse alpha 2-macroglobulinlike antigenic determinants on *Schistosoma mansoni* adults, *J Parasitol* **62**:413-419.

105. McLaren D.J., 1984, Disguise as an evasive stratagem of parasitic organisms, *Parasitology* **88**:597-611.

106. Johnson P., Garland, P.B., Campbell, P., *et al.*, 1982, Changes in the properties of the surface membrane of *Schistosoma mansoni* during growth as measured by fluorescence recovery after photobleaching, *FEBS Lett* **141**:132-135.

107. Foley M., Macgregor, A.N., Kusel, J.R., *et al.*, 1986, The lateral diffusion of lipid probes in the surface membrane of *Schistosoma mansoni*, *J Cell Biol* **103**:807-818.

108. Redman C.A. and Kusel, J.R., 1996, Distribution and biophysical properties of fluorescent lipids on the surface of adult *Schistosoma mansoni*, *Parasitology* **113**:137-143.

Chapter 7

THE APPLICATION OF DNA MICROARRAYS IN THE FUNCTIONAL GENOMIC STUDY OF SCHISTOSOME/HOST BIOLOGY

Karl F. Hoffmann and Jennifer M. Fitzpatrick
Department of Pathology, University of Cambridge, Tennis Court Road, CB2 1QP, UK

Key words: DNA microarrays, immunopathology, gender-specific genes, expression

1.   INTRODUCTION

The end of the 20th century and the start of the new millennium have bore witness to a remarkable revolution in the way parasite/host biological interactions can be conceptually designed and experimentally studied. Although most traditional investigations of parasitism have been motivated by hypothesis- or model-driven science, the recent successes of parasite and host expressed sequence tag (EST)/genomic sequencing projects have opened up a new avenue to the parasitologist – discovery driven science. Even the most steadfast, classically-trained parasitologist can see enormous value in coupling conventional experimental techniques with the new and exciting technologies made possible by genome sequencing efforts. In this era of functional genomics (defined as experimental approaches that use genomic structural information to understand biology in a systemic and comprehensive fashion (1)) the genome-wide analysis of mRNA expression using DNA microarrays has become pivotal. Here, we review the impact DNA microarrays have had on recent schistosome/host investigations and outline the exciting future areas we plan to visit in our study of schistosome sexual maturation, developmental biology, gender interactions, and host immuno-biology.

2.   DNA MICROARRAYS

Use of glass slide DNA microarrays to study transcriptional changes between two different nucleic acid samples obtained from eukaryotes was first reported in 1995-1996, when Patrick Brown's group at Stanford University published a series of seminal observations on *Arabidopsis* and *Saccharomyces* gene expression (2, 3). These high profile investigations brought DNA microarray technology worldwide publicity, building on the principles first conceived by Ekins and Chu during their development of solid-support immunoassays in the 1980's (4). DNA microarray technology has exploded since these early reports of its usage and all aspects of this evolving functional genomics tool have recently been critically re-reviewed in Nature Genetics (Supplement Volume 32, 2002).

A major advantage of utilizing DNA microarrays to study gene expression in the functional genomics era is that current technological improvements in robotics have made it possible to systematically array thousands of DNA elements onto chemically defined matrixes. In some cases, whole genomes of pathogenic organisms can be represented on a single glass slide (5). This means that gene expression profiles of whole genomes can be assayed simultaneously, exponentially increasing the amount of biological information obtained every time a single experiment is performed. However, highly informative DNA microarray studies have already been performed on partial parasite genomes (eg. *Plasmodium* (6)); this will continue to be the trend especially for those parasite species where complete genome annotation is unlikely in the near future (eg. *Ascaris*, *Echinococcus*, and *Trichuris*).

The vast amounts of sequence information available for mouse and man has made DNA microarrays from these schistosome definitive hosts an additional attractive resource for numerous questions related to parasite/host interactions. These questions might include how immunopathology develops, how immune responses are induced and regulated during infection, and what immune correlates are associated with resistance and vaccine induced protection. Together, the combined use of schistosome and host DNA microarrays in modern parasitological experimentation will be instrumental in generating comprehensive views of the mechanisms influencing host/parasite immuno-biology.

Two of our recent studies utilizing DNA microarrays have provided insight into the immunopathology of murine schistosomiasis (7) as well as the sexual dichotomy of adult parasites (8). As these two studies provided the framework for our continued interest in using DNA microarrays for future investigations into schistosome biology, they will be discussed below with attention given to the major findings included in each investigation.

3. STUDYING EGG-INDUCED IMMUNO-PATHOLOGY IN MURINE SCHISTOSOMIASIS USING cDNA MICROARRAYS

The schistosome egg is the causative agent that precipitates most severe forms of immunopathology in the murine model of schistosomiasis (see chapters 8 and 9). Distinct forms of clinical disease exist in the mouse model and these differences are attributable to the specific type of immune reactions developing in response to schistosome eggs (recently reviewed (9)). For example, schistosome-infected interleukin 10 (IL-10)/IL-12 (type-2 polarized) and IL-10/IL-4 (type-1 polarized) – deficient mice display distinct, and non-overlapping egg-induced pathologies, yet both groups of animals suffer from elevated mortality rates (10) in comparison to infected WT (wild-type) mice. Type-1 polarized animals display 100% mortality at 8 weeks post-infection and exhibit elevated splenic iNOS activity, increased production of pro-inflammatory cytokines (found in both liver and serum), decreased hepatic fibrosis, increased levels in circulating liver transaminase, and rapid cachexia. In contrast, the type-2 polarized animals develop a progressive wasting disease that correlates with increased production of the pro-fibrotic cytokines IL-4 and IL-13, augmented hepatic fibrosis, and significant mortality (~ 50%) during the chronic stages of infection. For greater insight into the genetic differences associating with these diverse egg-induced pathologies, a murine cDNA microarray fabricated from 2200 unique DNA elements was used to obtain gene expression profiles from hepatic tissue harvested from both infected mouse strains (7). Profiles obtained for the infected type-1 polarized mice illustrate that aberrant regulation of the apoptotic machinery, increased recruitment of activated macrophages and neutrophils, and diminished ability to transcribe genes associated with collagen production are likely contributing to the egg-induced hepatic pathology. In contrast, hepatic transcriptional analysis of the infected type-2 polarized animals clearly demonstrate that genes involved in wound repair, extra-cellular matrix remodeling, and collagen deposition are induced to a much greater degree than observed for the type-1 polarized mice. These findings thus provided a further mechanistic explanation for the increased mortality rates in type-2 polarized mice and illustrate that type-2 responses promote and type-1 responses inhibit gene transcriptional pathways associated with liver collagen synthesis and matrix remodeling. Furthermore, this study produced a list of candidate genes involved in egg-induced pathology and details how gene expression profiles can lead to insight (role of apoptosis, activated neutrophils and macrophages) into previously unappreciated disease mechanisms that contribute to pathogenesis during infection with schistosomes.

4.   STUDYING SCHISTOSOME SEXUAL MATURATION USING cDNA MICROARRAYS

Shortly after the above immunopathological study, we became interested in using DNA microarrays to examine schistosome sexual maturation and developmental biology.   The two experimental disciplines are logically linked, as murine pathology can be directly related to the biology of parasite sexual maturation.   Specifically, adult male schistosomes induce sexual maturation in females (11), which leads to the production of the 100-300 eggs/worm pair every day during infection with this trematode.   Any additional information relating to the processes regulating or associating with schistosome sexual maturation and development can be instrumental in understanding, and eventually controlling the immunopathology of disease. In a preliminary experiment aimed at identifying novel sex-associated gene transcripts, we fabricated a small *S. mansoni* cDNA microarray from three cDNA libraries originating from two different parasite developmental stages and probed these DNA microarrays with sexually mature male and female material (8).   Although the fabricated cDNA microarray consisted of less than 10% of the predicted ORF content of the *S. mansoni* genome, this study expanded the list of gender-associated gene transcripts from all previous studies by a factor of two and provided novel research avenues towards our understanding of the molecules involved in egg-production, tegumental biology, and other sex-related biological processes.   We, and others, are in the process of functionally characterizing the biological importance of many of these differentially expressed transcripts.

Due to this pilot experiment's success, we have recently begun the fabrication of an expanded *S. mansoni* oligonucleotide DNA microarray (Fig 1, panel I - containing ~ 50% of the predicted ORF genome content) as well as a small *S. japonicum* cDNA microarray (Fig 1, panel V - 456 unique cDNA clones or ~ 3% of the predicted ORF genome content). These two resources will be extensively used over the next several years to aid our functional genomic research of schistosome biology (discussed in greater detail below).

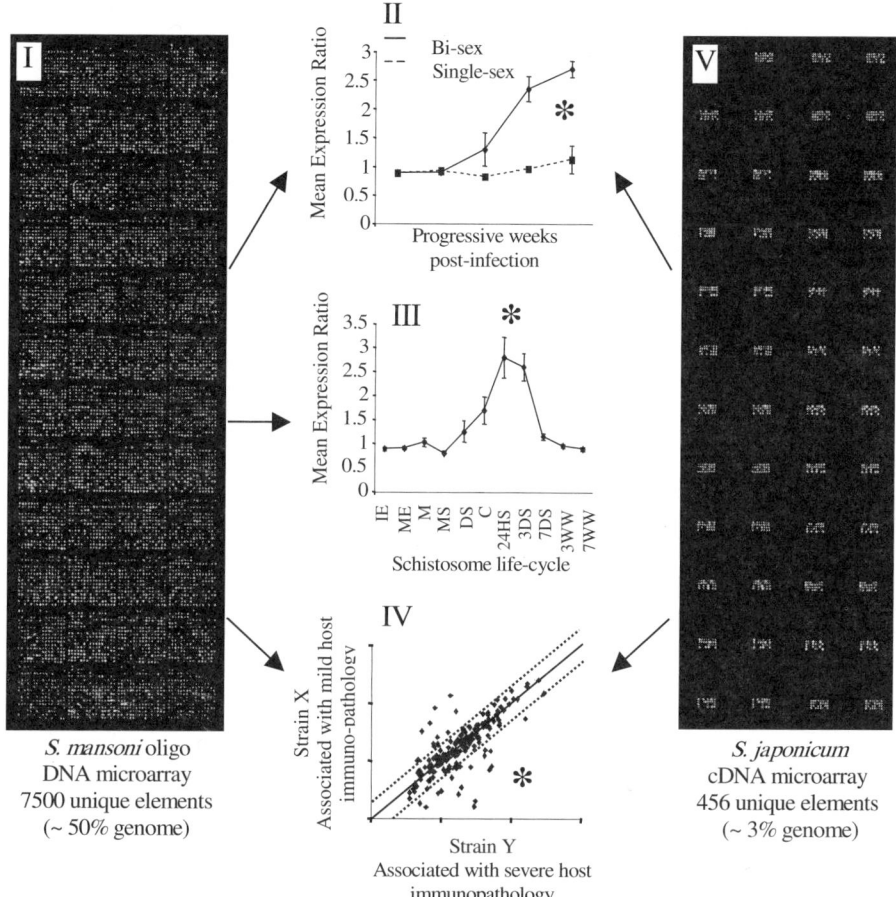

Figure 7-1. Schistosome specific DNA microarrays that will be used in future functional genomic studies of host-schistosome interactions. (*S. mansoni* oligonucleotide DNA microarrays containing 7500 unique elements (I) and *S. japonicum* cDNA microarrays containing 456 unique elements (V)) will be used to identify transcripts associated with: (II) sexual maturation, (III) developmental expression, and (IV) immunopathology. (II) An example of one planned sexual maturation study result where * indicates average gene expression profile of transcripts only induced in sexually mature female schistosomes obtained from bi-sex infection experiments. (III) An example of one planned developmental expression profiling result where * indicates average gene expression of transcripts highly induced in 3 day through 7 day schistosomula. IE – immature egg, ME – mature egg, M – miracidia, MS – mother sporocyst, DS, daughter sporocyst, C – cercaria, 24HS – 24 hr schistosomula, 3DS – 3 day schistosomula, 7DS – 7 day schistosomula, 3WW – 3 week sexually immature worms, 7WW – 7 week sexually mature worms. (IV) Scatterplot of gene expression ratios obtained from 2 schistosome strains capable of inducing diverse immunopathology; * indicates those genes associated with the more virulent strain of parasite.

5.   WHAT'S NEXT AND EXCITING FUTURE PROSPECTS

Based on the above-described studies, we are excited about this technology's impact on future parasitological investigations. We plan on using parasite DNA microarrays to 1) continue studies into the sexual biology and developmental maturation of schistosomes, 2) investigate whether parasite gene expression profiles can be predictive of virulence and/or immunopathology, and 3) identify potential new immuno-prophylactic or chemotherapeutic targets. In addition to these investigations using parasite DNA microarrays, we also are currently using both human and murine DNA microarrays to study immune response induction and regulation during praziquantel chemotherapy of infected hosts. Transcriptional results obtained from these studies, in combination with co-acquired immunological and parasitological data, will shed light into the mechanisms associating with host resistance, susceptibility, and the generation of a dominant and controlled type-2 response. The interpretation of data collected from both parasite and host DNA microarray studies will be instrumental in developing new hypotheses that aid in the characterization of complex, infection-associated, biological interactions.

6.   FUTURE USE OF SCHISTOSOME DNA MICROARRAYS

The goals of our future sex-associated studies are to use the above-described schistosome DNA microarrays to identify: 1) genes transcribed in sexually mature adult male and female parasites, 2) genes longitudinally transcribed during sexual development and maturation, and 3) genes differentially regulated in murine infections initiated by single sex versus dual-sex parasite inoculums. In addition to expanding the list of transcripts that associate with adult *S. mansoni* males and females, we plan on identifying adult gender-associated gene products from *S. japonicum* worms, where even less is known about its sexual maturation and egg biology. We therefore hope to contribute to the identification of sex-associated gene transcripts in both species of schistosome.

Although bimodal (adult male versus female) experiments will lead to the cataloguing of many additional adult gender-associated transcripts, we are equally interested in identifying those genes that are longitudinally regulated from immature through mature forms of these dioescious parasites. The timing of gene expression will provide important clues into the development of sexual maturity, conjugal biology, oviposition, and host/parasite interactions (12). Gene transcripts that demonstrate female associations in the bimodal experiment described above can be additionally segregated into those that are temporally expressed before and after egg development. Therefore, male and female gene expression profiles will be obtained from worms harvested from infected murine hosts at progressive weeks post-infection and probed across the DNA microarrays. These experiments will help elucidate the transcriptional events associated with the acquisition of schistosome sexual maturity.

Although it is envisioned that sex-associated transcription kinetics will be characterized in the above experiments, a third biological area to be addressed during our studies of schistosome sexual biology will involve identifying those genes differentially regulated in single-sex (hosts only infected with one parasite gender) versus dual-sex (hosts infected with both parasite genders) murine infections (*, Fig. 1, panel II). It has been shown that female parasites depend upon direct interaction with male schistosomes to mature into a highly proficient egg-laying organism (13). Conversely, the male schistosome seems capable of normal sexual development in the absence of the female, but some studies suggest that antigenic and biochemical changes do occur (14, 15). In both cases, the genetic machinery affected by maturing in the absence of the opposite sex has yet to be thoroughly dissected. To address this important subject, female and male parasites from single sex infections of each (16) will be harvested from murine hosts at progressive weeks post-infection. The longitudinal pattern of gene expression will be assessed from each separate sex pool of RNA and then compared to those pools of separate sex RNA harvested from a dual sex murine infection via a common reference. It is anticipated that these experiments will uncover contributions of previously unappreciated mechanisms involved in schistosome sexual maturation pathways, which depend upon gender interactions.

To complement these studies of schistosome sexual biology, we plan to use the *S. mansoni* DNA microarrays to profile gene expression across the parasite's lifecycle and identify gene products associated with different parasite strains. The goal of lifecycle expression profiling is to identify coordinated clusters of gene transcripts that associate with each particular life-stage. Using this approach, we will be able to characterize the developmental biology of this pathogen to a large degree, which ultimately will identify transcripts involved in housekeeping functions, uniquely utilized metabolic pathways, infection of intermediate and definitive hosts, and modulation of protective immune mechanisms (Fig. 1, panel III). Further detailed elucidation of these genes' functions may provide insight into novel strategies for immunological or drug intervention.

Finally, we plan to use both *S. mansoni* and *S. japonicum* DNA microarrays as tools to profile gene transcription differences and similarities among various collected field isolates (strains) and laboratory adapted parasite strains. Information obtained from these investigations, in association with other field- and laboratory- collected parameters, may provide significant information related to the virulence of the profiled parasite strains useful in interpreting complicated immuno-epidemiological associations (*, Fig. 1, panel IV). Taken together, data accumulated from our future use of schistosome DNA microarrays will provide novel, testable hypotheses related to diverse biological processes concerning host-parasite interactions, developmental biology, virulence, and sexual maturation of these pathogenic organisms.

7.   FUTURE USE OF HUMAN AND MOUSE DNA MICROARRAYS

Investigations using host specific DNA microarrays are also currently being used in our laboratory to study immune response regulation during chemotherapeutic treatment. Because any vaccine administered to a human population will likely be preceded by chemotherapeutic clearance of an active infection, understanding the effect of drug therapy on the host's immune system is an important issue to investigate and will ultimately contribute to the development of an effective immunoprophylactic treatment. The present drug of choice for treatment of all human schistosomes is praziquantel. While the exact mode of action of praziquantel is still not clear, experimental mouse studies (17-19) supported by human investigations (20), suggest that praziquantel therapy induces a protective immune response. However, according to human re-infection studies, this protective immune response may not be long-lived in many individuals (e.g. children have increased risk of becoming re-infected) (21, 22). Consequently, characterization of the immunological mechanisms induced by praziquantel treatment that contribute to worm elimination should be more carefully and thoroughly examined. A study of this type would not only provide information critical to schistosome vaccinologists, but would also furnish the immunologist with insights into how the immune system regulates a potentially deleterious response against thousands of parasite antigens that are simultaneously released into the blood at the time of chemotherapy.

Studies to date dissecting the human immune responses longitudinally after praziquantel treatment have been limited. Most have examined the effect of chemotherapy on some aspect of the host immune system (using sera, PBMCs, or whole blood assays) at periods of several weeks (20, 23), months (22, 24, 25), or years (21, 25) after treatment. Although the majority of these studies have revealed the up-regulation of IgE and IgG responses as well as increased eosinophilia after praziquantel treatment, all of them have neglected to examine the early events induced immediately following chemotherapy. Speculatively, these early host immuno-biological responses to dying worms could directly influence the magnitude of later events including the development of heightened antibody titers, proliferation of eosinophils, and ultimately, generation of immunity to a subsequent infection.

Our current studies involve examining the host's transcriptional profile longitudinally during praziquantel treatment, focusing on immediate as well as late time-points (Fig. 2). In the murine studies, where careful control of the experimental procedure can be achieved, we are currently examining splenic, bonemarrow, and mesenteric lymph node expression profiles during praziquantel treatment in both infected as well as uninfected animals. A DNA microarray consisting of 7445 unique elements (~33% of the predicted ORF content of the genome) will be utilized for these studies in the mouse (Fig. 2 panel III). Our human studies focus on a well-characterized cohort of individuals living on Lake Albert in Uganda, Africa. Adults living in this area are all infected with *S. mansoni* and have previously been shown to homogenously mount a rapid and vigorous immune response against parasite antigens following chemotherapy. Because of a high infection rate and rapid post-treatment immune response, these individuals provide a unique opportunity to thoroughly characterize the extent, quality, and magnitude of gene activity induced or repressed by chemotherapy and dissect how these differentially regulated genes influence the development of later immunological events (*, Fig. 2, panel II). Here, we are using a DNA microarray (Fig. 2, panel I) consisting of 10560 unique elements (~33% of the predicted ORF content of the genome) and are specifically focusing on gene expression profiles obtained from whole blood (where the red blood cell component has been effectively removed). The contribution of information collected from both murine and human experiments via DNA microarray analysis of selected tissues, in combination with classical immunological and parasitological studies, will help build a more robust picture of the immunological cascade triggered by praziquantel treatment of infected and uninfected individuals. This information should prove useful towards our advanced understanding of praziquantel's protective mechanism of action as well as provide important clues to the gene products that may be associated with resistance to re-infection; both topics remain a priority among planned future immunological investigations.

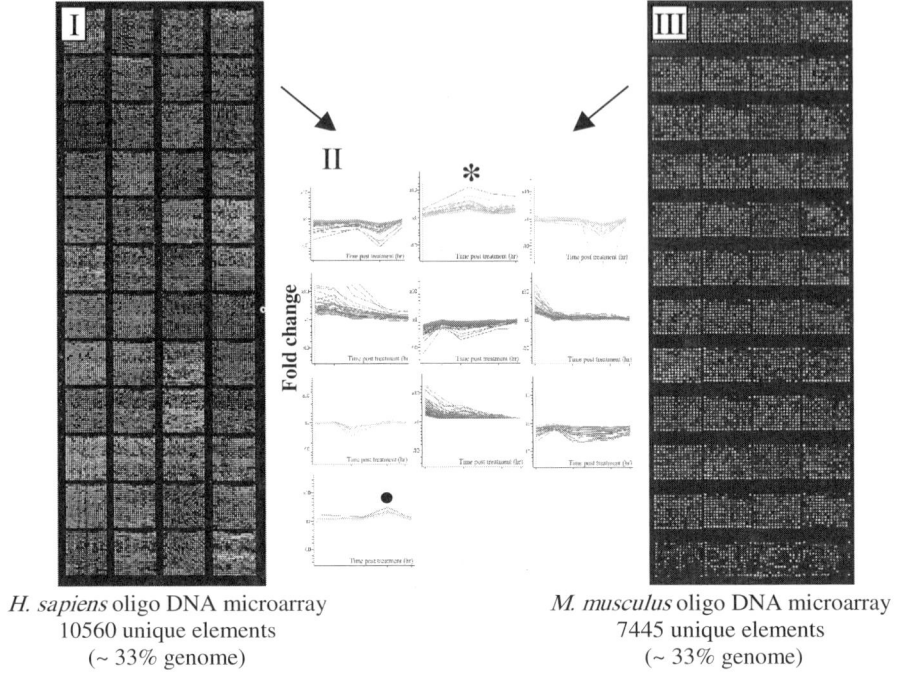

H. sapiens oligo DNA microarray
10560 unique elements
(~ 33% genome)

M. musculus oligo DNA microarray
7445 unique elements
(~ 33% genome)

Figure 7-2. Host specific DNA microarrays (*H. sapiens* oligonucleotide DNA microarrays containing 10560 unique DNA elements (I) and *M. musculus* oligonucleotide DNA microarrays containing 7445 unique DNA elements (III)) will be used as a tool to identify patterns of gene expression differentially regulated upon praziquantel clearance of an active schistosome infection. (II) K-means supervised clustering method used to segregate gene expression profiles differentially regulated during praziquantel treatment of schistosome-infected hosts. * Indicates genes induced early after praziquantel treatment and • indicates genes induced later after drug therapy (possible effector or regulatory transcripts). Other patterns of gene expression influenced by drug therapy can be visualized in the additional clusters.

8.   CONCLUDING REMARKS

Although gene expression results obtained through the use of DNA microarrays will undoubtedly increase the path and scope of many future parasite/host investigations, there is still an obvious need to maintain the use of traditional experimental techniques in this functional genomics age. How else would the gene products identified by DNA microarray investigations be further studied in the laboratory setting if we do not plan on using this wealth of information for testing hypotheses by current immunological, biochemical, and molecular biological techniques? In addition to classical techniques, new methodologies can be used to manipulate parasite and host gene expression. Transgenic or gene deficient cell lines or animals are well established and RNAi and transient/stable transfection of parasites and host cells are being developed which collectively allow for the functional importance of selected gene products to be determined directly.

Each investigator considering the use of DNA microarrays will additionally face the rewarding task of sifting through huge amounts of gene expression information to tease apart biologically significant relationships. The evolution of new bioinformatic approaches for DNA microarray data analysis will dramatically increase the speed in which these biological relationships are identified as well as provide novel opportunities for developing collaborative relationships between the biologist and the mathematician. In terms of future research on schistosomiasis, DNA microarrays and other functional genomic tools have the profound potential to positively impact all areas, provided widespread use is routinely adopted alongside our traditional immuno-parasitological methodologies.

REFERENCES

1. Staudt, L. M., and P. O. Brown. 2000. Genomic views of the immune system*. *Annu Rev Immunol 18:829.*
2. Schena, M., D. Shalon, R. W. Davis, and P. O. Brown. 1995. Quantitative monitoring of gene expression patterns with a complementary DNA microarray. *Science 270:467.*
3. Shalon, D., S. J. Smith, and P. O. Brown. 1996. A DNA microarray system for analyzing complex DNA samples using two- color fluorescent probe hybridization. *Genome Res 6:639.*
4. Ekins, R., and F. W. Chu. 1999. Microarrays: their origins and applications. *Trends Biotechnol 17:217.*
5. Fitzgerald, J. R., and J. M. Musser. 2001. Evolutionary genomics of pathogenic bacteria. *Trends Microbiol 9:547.*
6. Hayward, R. E., J. L. Derisi, S. Alfadhli, D. C. Kaslow, P. O. Brown, and P. K. Rathod. 2000. Shotgun DNA microarrays and stage-specific gene expression in *Plasmodium falciparum* malaria. *Mol Microbiol 35:6.*
7. Hoffmann, K. F., T. C. McCarty, D. H. Segal, M. Chiaramonte, M. Hesse, E. M. Davis, A. W. Cheever, P. S. Meltzer, H. C. Morse, 3rd, and T. A. Wynn. 2001. Disease fingerprinting with cDNA microarrays reveals distinct gene expression profiles in lethal type 1 and type 2 cytokine-mediated inflammatory reactions. *Faseb J 15:2545.*
8. Hoffmann, K. F., D. A. Johnston, and D. W. Dunne. 2002. Identification of *Schistosoma mansoni* gender-associated gene transcripts by cDNA microarray profiling. *Genome Biol 3:RESEARCH0041.*
9. Hoffmann, K. F., T. A. Wynn, and D. W. Dunne. 2002. Cytokine-mediated host responses during schistosome infections; walking the fine line between immunological control and immunopathology. *Adv Parasitol 52:265.*

10. Hoffmann, K. F., A. W. Cheever, and T. A. Wynn. 2000. IL-10 and the dangers of immune polarization: excessive type 1 and type 2 cytokine responses induce distinct forms of lethal immunopathology in murine schistosomiasis. *J Immunol 164:6406.*
11. Erasmus, D. A. 1973. A comparative study of the reproductive system of mature, immature and "unisexual" female *Schistosoma mansoni. Parasitology 67:165.*
12. Braga, V. M., C. A. Tavares, and F. D. Rumjanek. 1989. Protein characterization of sexually mature and immature forms of *Schistosoma mansoni. Comp Biochem Physiol B 94:427.*
13. Popiel, I., D. Cioli, and D. A. Erasmus. 1984. The morphology and reproductive status of female *Schistosoma mansoni* following separation from male worms. *Int J Parasitol 14:183.*
14. Cornford, E. M. 1985. *Schistosoma mansoni, S. japonicum,* and *S. haematobium:* permeability to acidic amino acids and effect of separated and unseparated adults. *Exp Parasitol 59:355.*
15. Aronstein, W. S., and M. Strand. 1983. Identification of species-specific and gender-specific proteins and glycoproteins of three human schistosomes. *J Parasitol 69:1006.*
16. Shaw, J. R., and D. A. Erasmus. 1981. *Schistosoma mansoni:* an examination of the reproductive status of females from single sex infections. *Parasitology 82:121.*
17. Sabah, A. A., C. Fletcher, G. Webbe, and M. J. Doenhoff. 1985. *Schistosoma mansoni:* reduced efficacy of chemotherapy in infected T-cell-deprived mice. *Exp Parasitol 60:348.*
18. Andrews, P. 1985. Praziquantel: mechanisms of anti-schistosomal activity. *Pharmacol Ther 29:129.*
19. Shaw, M. K. 1990. *Schistosoma mansoni:* stage-dependent damage after in vivo treatment with praziquantel. *Parasitology 100 Pt 1:65.*
20. Mutapi, F., P. D. Ndhlovu, P. Hagan, J. T. Spicer, T. Mduluza, C. M. Turner, S. K. Chandiwana, and M. E. Woolhouse. 1998. Chemotherapy accelerates the development of acquired immune responses to *Schistosoma haematobium* infection. *J Infect Dis 178:289.*
21. Satti, M. Z., P. Lind, B. J. Vennervald, S. M. Sulaiman, A. A. Daffalla, and H. W. Ghalib. 1996. Specific immunoglobulin measurements related to exposure and resistance to *Schistosoma mansoni* infection in Sudanese canal cleaners. *Clin Exp Immunol 106:45.*
22. Kabatereine, N. B., B. J. Vennervald, J. H. Ouma, J. Kemijumbi, A. E. Butterworth, D. W. Dunne, and A. J. Fulford. 1999. Adult resistance to schistosomiasis mansoni: age-dependence of reinfection remains constant in communities with diverse exposure patterns. *Parasitology 118:101.*
23. Butterworth, A. E., P. R. Dalton, D. W. Dunne, M. Mugambi, J. H. Ouma, B. A. Richardson, T. K. Siongok, and R. F. Sturrock. 1984. Immunity after treatment of human schistosomiasis mansoni. I. Study design, pretreatment observations and the results of treatment. *Trans R Soc Trop Med Hyg 78:108.*
24. Naus, C. W., G. J. van Dam, P. G. Kremsner, F. W. Krijger, and A. M. Deelder. 1998. Human IgE, IgG subclass, and IgM responses to worm and egg antigens in schistosomiasis haematobium: a 12-month study of reinfection in Cameroonian children. *Clin Infect Dis 26:1142.*
25. Roberts, S. M., R. A. Wilson, J. H. Ouma, H. C. Kariuki, D. Koech, T. K. arap Siongok, R. F. Sturrock, and A. E. Butterworth. 1987. Immunity after treatment of human schistosomiasis mansoni quantitative and qualitative antibody responses to tegumental membrane antigens prepared from adult worms. *Trans R Soc Trop Med Hyg 81:786.*

Chapter 8

THE INITIATION OF HOST IMMUNE RESPONSES TO SCHISTOSOME EGG ANTIGENS

Edward J. Pearce
University of Pennsylvania, Phildelphia, PA 19104-6076

Key words: Th1/Th2; helminth parasite; schistosoma; dendritic cell

1.   INTRODUCTION

An unusual feature that distinguishes schistosomiasis from many other diseases is that weeks into infection the parasites begin to produce a new life stage, the egg, that expresses a subset of genes that are not transcribed in the worm stages. Thus the host is suddenly exposed to new and novel antigens (Ag).   This transition is accompanied by two significant events: inflammation in the liver and intestine as the host responds to parasite eggs in these tissues, and the development of marked cellular and humoral responses against the novel egg Ag. Since the nature and intensity of the anti-egg immune response plays a role of central importance in dictating disease severity, there has been significant focus of investigation on the egg-induced immune response (1).

2.   CHANGES IN THE IMMUNE RESPONSE DURING THE PREPATENT-PATENT TRANSITION

During the pre-patent period of infection, during the first 4-5 weeks following exposure to cercariae, the immune response is primarily Th1 in nature and, as would be expected given the life stages the host is exposed to during this time, directly primarily against worm Ag (2, 3). However, as the first eggs are produced by the parasites the immune response takes on an entirely different character, becoming strongly Th2-polarized by 8 weeks post infection (2, 3). This Th2 response is egg Ag specific, and its development is accompanied by the loss of the worm Ag specific Th1 response that preceded it (3). Coincident with the development of the Th2 response, there is a dramatic increase in plasma IgE levels and circulating eosinophil numbers which reflect the production of IL-4 and IL-5, signature cytokines of Th2 cells, that respectively help B cells to class switch Ig isotype production to IgE and act as a growth and survival factor for eosinophils (4, 5). Later in mixed sex infection, after about 3 months, a significant diminution of the Th2 response is apparent and this state of comparative hyporesponsiveness persists for the remainder of the infection (2). In contrast, in animals infected with single sex cercariae, where egg production does not occur, the worm Ag specific Th1 response persists (2).

3.   Th2 RESPONSE INDUCTION IS AN INHERENT PROPERTY OF SCHISTOSOME EGGS

Eggs isolated from infected mice and injected into normal mice induce markedly Th2-polarized responses. This outcome occurs regardless of whether the eggs are injected intravenously (6), subcutaneously (7) or intraperitoneally (8), and is similar regardless of whether the eggs are alive or have been killed by freezing and thawing (3). The physical structure of the egg, acting as an Ag deport, contributes partially to the antigenicity of this life stage, but SEA, a PBS-soluble extract of homogenized eggs, is also able to induce Th2 responses when injected subcutaneously or administered intranasally (9) or intratracheally (10). Eggs and egg Ag induce Th2 responses independently of any requirement for a Th2 response promoting adjuvant such as Alum, and indeed can act as adjuvants themselves, driving responses to unrelated antigens in a Th2 direction (11). Thus schistosome egg Ag are unusually potent inducers of Th2 responses. There is strong evolutionary pressure to maintain this property, since the inability to make Th2 responses renders animals highly susceptible to developing life-threatening disease when infected with schistosomes (12-15).

SEA is a highly complex mixture of all of the PBS soluble proteins from schistosome eggs. Presumably this largely represents cytoplasmic proteins from the various cells within the egg, plus the proteins that are destined for secretion. In all probability it is the latter that are of most importance, since it is these that are released from living eggs and against which the cellular response is initiated during infection. Despite the importance of this set of

Ag, very few egg proteins have been immunologically characterized in any detail. Thus at this stage, although it is clear that certain egg proteins are quite antigenic (16), it is unlcear whether there is a dominant Th2 response eliciting egg Ag *per se*. However, it is becoming increasingly clear that unusual α3-fucose and β2-xylose containing sugar modifications of egg molecules play an important role in Th2 response induction (17) (18). Of importance amongst these is lacto-N-fucopentaose III, which contains the Lewis X trisaccharide. Remarkably this sugar confers Th2 response inducing properties to protein Ag to which it is coupled (11, 19). Moreover, egg Ag lose their ability to induce Th2 responses when chemically deglycosylated or periodate-treated (11).

4. Th RESPONSE POLARIZATION IS CONTROLLED BY A COMPLEX BALANCE OF FACTORS

Upon first exposure to Ag, previously inexperienced Th cells begin to produce IL-2, an autocrine growth factor, and enter the cell cycle. As the cells divide, they acquire the ability to synthesize additional cytokines (20). Thereafter, responses can polarize to become dominated by IFNγ producing Th1 cells, or IL-4/IL-5/IL-13 producing Th2 cells. Of great importance in this process are the cytokines IL-12 and IL-4 (20). IL-12 is made by cell types that can process and present Ag to Th cells, such as dendritic cells and macrophages, and additionally by cells such as neutrophils (21). The involvement of cells such as these in the innate response to Th1 response inducing pathogens during the period preceding the initiation of adaptive Th responses can create an environment rich in IL-12, which promotes the outgrowth of Th cells making IFNγ. IFNγ itself stabilizes the expression of the IL-12R on Th cells and in this way helps consolidate the Th1 response (20). Other cytokines such as IL-18 and IL-27 support this pathway (22). Thus the prevailing view of Th1 response development is one in which non Th cells that are involved in innate immunity produce factors that promote polarization of the adaptive T cell response.

For Th2 response development IL-4 rather than IL-12 is a crucial polarizing factor (20). Thus infection in mice that lack IL-4 fails to induce egg Ag specific Th2 responses (12, 23) whereas the absence of IL-12 has no effect on egg-induced Th2 response development (24). The cellular source of the IL-4 important for Th2 response development is generally unclear. Eggs and/or Ag released from them innately activate mast cells (25) and basophils (26, 27), cell types which are able to make IL-4. Moreover, as part of an early innate response, in mice that have been injected with eggs mast cells elicit the recruitment of IL-4 producing eosinophils (25). Nevertheless, eggs will induce Th2 responses of normal intensity in the absence of either mast cells or eosinophils (25, 28). Moreover, other cell types that can make IL-4 and have been implicated in Th2 response development to other pathogens, such as NK T cells, are not essential for Th2 response induction by injected eggs (29). Thus, it seems likely that, as was originally suggested almost a decade ago (30), the IL-4 necessary for Th2 response polarization is being provided by the Th cells themselves (20).

5.   HOW ARE ANTI-EGG IMMUNE RESPONSES INITIATED?

Immune responses are generally held to be initiated by dendritic cells (DC), specialized highly phagocytic cells that sample their environment and proteolytically degrade acquired proteins into peptides that can be complexed with MHC class II molecules for display at the cell surface (31). These peptide/MHC class II complexes are the ligands recognized by CD4 T cell Ag receptors. DC are present throughout the body, especially at the epithelial interfaces with the exterior through which natural exposure to pathogens usually occur. As a result of the coordinated expression of chemokine receptors DC are able to home from these peripheral sites to the T cell areas of lymphoid organs, where they can meet T cells of the appropriate specificity. T cell activation by DC requires costimulatory signals in addition to those delivered through the TCR. These are provided by molecules such as CD80 and CD86 that are expressed at the DC surface and ligate CD28 on T cells. Moreover, DC are a potent source of IL-12 and thus can play a pivotal role in polarizing Th responses in a Th1 direction (21). Increased expression of peptide/MHC class II and molecules such as CD80 and CD86 and changes in chemokine receptor and cytokine/chemokine production are but a few of the changes that occur to DC as they undergo the activation process referred to as maturation. While macrophages and B lymphocytes also can express MHC class II and costimulatory molecules, they are notably less capable than DC of providing the necessary signals to activate naïve Th cells.

DC maturation can be induced by immune system molecules such as IFNαβ, TNFα• or CD40L, or by ligation of toll like receptors (TLR) (32). Mammalian genomes encode •10 TLR, each of which is able to bind to distinct pathogen associated molecular patterns (PAMPs). In this way, cells such as DC can innately recognize as foreign a large array of microbial pathogens. It is possible that schistosomes do not make any molecules that are recognized as PAMPs by TLR since, unlike many microbial pathogens, schistosome egg Ag fail to directly induce DC maturation (33, 34).

6.   WHAT EFFECT DOES SEA HAVE ON DC?

SEA-loaded DC injected either intraperitoneally or subcutaneously into naïve recipient mice induce the development of egg Ag-specific Th2 responses (33, 35). In these experiments the recipient mice were presumably not exposed to any SEA components apart from the peptides displayed on the surface of the DC. The important implication of this finding is that DC alone are sufficient to elicit SEA-specific Th2 responses – there is no requirement for SEA to interact with other cell types for this type of response to develop. Thus further analysis of the interaction between DC and SEA should hold the answer to how SEA induces a Th2 response. However, detailed analyses of bone marrow derived DC cultured in the presence of SEA have to date failed to reveal any major direct transcriptional response of the cells to the Ag (J. Sun, A. Straw and E. Pearce, unpublished). Thus there is currently little evidence for an SEA-induced alternative pathway of activation or maturation that could be linked to Th2 response development. Nevertheless, this possibility remains appealing.

The lack of an easily detectable response of DC to SEA has led to the idea that SEA induced Th2 responses represent a default that occurs in the absence of any overt DC activation. Put simply, SEA-specific Th2 responses may develop simply because SEA fails to elicit IL-12 production. It is clear that the presence of this cytokine can have a dramatic effect on SEA induced immune responses, since coinjection of eggs with IL-12 (36), leads to the development of an SEA-specific Th1 response. Thus exogenous IL-12 is sufficient to sway the balance of the egg Ag specific response. Recent evidence indicates that SEA can prevent DC from making IL-12 in response to other inflammatory stimuli such as LPS (37) (L. Cervi, C. Kane and E. Pearce, unpublished) a finding that it is tempting to believe is related to the Th2 response inducing properties of this helminth Ag.

An alternative possibility is that it is some aspect of DC maturation unrelated to IL-12 that is of greatest importance for Th1 response induction and that it is the absence of this other factor that allows SEA-pulsed DC to induce Th2 responses (1). One of the most compelling arguments for this possibility has come from the study of the immune response to another parasite, *Toxoplasma gondii*. This parasite is noted for its ability to induce IL-12 production by DC, a process we now know to involve the cross-linking of a TLR and CCR5 by a parasite product that contains cyclophilin-18, and to induce highly Th1 polarized responses(38, 39). However, IL-12-/-

mice are able to continue to make Th1 responses to the parasite, but mice that lack MyD88, a key downstream component of the majority of TLR initiated signaling pathways, make Th2 responses when infected (40). Thus there appears to be a link between DC activation and Th1 response initiation. What the important non-IL-12 factor is in this process is not clear. One possibility is that activated DC present more peptide/MHC class II complexes and therefore a stronger activation signal to T cells with which they interact. High effective Ag dose/peptide density has been shown to be important for Th1 response induction in some model systems (41, 42).

DC pulsed with SEA do not make IL-4, and IL-4-/- DC are as effective as WT DC at inducing SEA specific Th2 responses (43). However, IL-4-/- recipients of SEA-pulsed WT DC do not make Th2 responses (43). As discussed above, Th cells themselves may be the important source of IL-4 necessary for Th2 polarization. However, recent studies have shown that DC pulsed with SEA are able to present a carbohydrate epitope (probably part of a glycolipid) complexed with CDd1 to activate NKT cells (35). The importance of these cells in Th2 response induction by SEA-pulsed DC was indicated by the fact that CD1-/- DC were unable to induce SEA-specific Th2 responses (35). Thus a situation can be envisaged in which SEA pulsed DC activate both Th cells and NKT cells and the IL-4 produced by the latter promotes the outgrowth of a Th2 response. While this is an attractive model, it does not account for the findings published earlier that injected eggs directly induce Th2 responses in β2m-/- mice, which do not express CD1 and which generally lack NKT cells, although it is possible that residual population of CD1d-restricted T cells remain in these animals,

7. DOES IN VIVO EVIDENCE SUPPORT THE VIEW THAT DC ARE NOT ACTIVATED BY SCHISTOSOME INFECTION?

The current paradigm in immunology is that immature DC are incapable of activating Th cells but rather induce tolerance by providing an incomplete activation signal (44). One implication of this is that SEA-pulsed DC must receive some type of maturation signal in order to induce Th2 responses rather than tolerance to SEA. The experimental evidence suggests that this signal is delivered by the ligation of CD40 on DC by CD40L. CD40-/- DC are incapable of initiating SEA specific Th2 responses but are competent to induce Th1 responses against bacteria (45). The importance of this DC activation pathway is reflected in the fact that CD40-/- and CD40L-/- mice are fatally immunodeficient when infected with *S. mansoni* (46).

DC isolated from the spleens or mesenteric LN of infected mice exhibit signs of maturation that are not apparent in SEA-pulsed bone marrow culture derived-DC (47). However, this state of maturation is not apparent in DC from infected CD40L-/- mice (47), further indicating that CD40 ligation is serving as an important immune system intrinsic signal for DC maturation during the initiation of Th2 responses to SEA. An intriguing facet of this activation pathway is that it can promote IL-12 production, and while SEA-pulsed bone marrow DC clearly do not make IL-12 following CD40 ligation (33), DC isolated from the spleens of infected mice are primed to make this cytokine and indeed produce more of it in response to CD40 ligation than do DC from uninfected mice (47). Splenic DC comprise several distinct subsets of DC, only one of which is analogous to the CD8α- bone marrow DC most extensively studied in schistosome research (32). Compared to these cells, CD8α+ DC are more potent IL-12 producers and it is likely that in the spleens of infected mice it is this subset that is capable of making IL-12. These observations raise several important questions: 1) Do CD8α+ DC and CD8α- DC respond differently to SEA? 2) Are DC in infected mice primed by schistosome Ag or immune system intrinsic signals to be able to make more IL-12 in response to CD40 ligation? 3) Perhaps most importantly, if during infection DC are primed to produce IL-12 in response to CD40 ligation, and are being induced to mature via this pathway, why does the immune response polarize in a Th2 rather than Th1 direction?

8.   A ROLE FOR IL-10

The answer to the third question posed in the preceding paragraph appears to be that Ag-responsive cells within the spleen make IL-10 which strongly inhibits production of IL-12 by DC (47) (A. Straw and E. Pearce, unpublished). IL-10 plays a crucial role regulating immune responses during infection (15). In its absence, highly pathogenic SEA specific Th1 responses develop in addition to the usual Th2 responses (14, 48). This is also the case in egg-injected mice, where again the absence of IL-10 allows Th1 response development, but interestingly is not the case in IL-10-/- mice injected with SEA-pulsed bone marrow DC (MacDonald and Pearce, unpublished). Thus while IL-10 is not necessary for the extreme Th2 polarization induced by SEA-pulsed bone marrow DC, it is essential to allow Th2 polarization when immune responses are developing during infection. A likely interpretation of these apparent discrepancies is that in infected and in egg-injected mice CD8α+ DC are being primed by inflammatory mediators in such a way that they produce IL-12 in response to CD40 ligation and possibly other signals, and that the production of IL-12 is strictly suppressed by IL-10 (47). At the time of writing there is considerable interest in the source of this IL-10, with evidence pointing strongly towards a major role for regulatory CD25+ T cells (49) (A. Straw and E. Pearce, unpublished). Interestingly in this regard, schistosome eggs have recently been reported to contain a lipid that is able to condition DC to promote T regulatory cell development (50).

9.   WHY IS SEA SO ANTIGENIC?

As intimated early in this chapter, SEA is a potent Ag, able to induce marked immune responses in the absence of adjuvant. One possibility is that SEA is very efficiently targeted to DC. It seems likely now that this is the case, since SEA has been shown to bind to DC-SIGN (51), a member of the large C-type lectin family, that acts as a pattern recognition receptor through which DC are able to recognize and take up foreign glycoconjugates. Targeting Ag to C-type lectin mediated uptake pathways has been shown to be a potent means of eliciting immune responses to otherwise weakly antigenic molecules (52). Whether utilization of the DC-SIGN pathway in any way contributes to the Th2-polarized nature of the SEA induced response is of considerable interest. Recent reports have indicated that ligation of DC-SIGN by *Mycobacterium tuberculosis* is an immune evasion pathway that results in a significant inhibition of the ability of DC to make IL-12 in response to TLR ligands expressed by these bacteria (53). The recent report that SEA pulsed DC are significantly less capable than non-pulsed DC of making IL-12 in response to TLR ligands (37) suggests that SEA too may be inhibiting DC function by binding to DC-SIGN .

ACKNOWLEDGEMENTS

The author wishes to thank Laura Cervi, Melinda Ekkens, Terry Fang, Colleen Kane, Andrew MacDonald, Amy Straw and Jie Sun for many helpful discussions and insights. Work in the author's laboratory is funded by grants from the NIH and the Ellison Medical Foundation. EJP is a recipient of a Burroughs Wellcome Fund Scholar in Molecular Parasitology award.

REFERENCES

1. Pearce, E. J., and A. S. MacDonald, 2002, The immunobiology of schistosomiasis. *Nat. Rev. Immunol.* **2**:499.
2. Grzych, J. M., E. Pearce, A. Cheever, Z. A. Caulada, P. Caspar, S. Heiny, F. Lewis, and A. Sher, 1991, Egg deposition is the major stimulus for the production of Th2 cytokines in murine schistosomiasis mansoni, *J. Immunol.* **146**:1322.
3. Pearce, E. J., P. Caspar, J. M. Grzych, F. A. Lewis, and A. Sher, 1991, Downregulation of Th1 cytokine production accompanies induction of Th2 responses by a parasitic helminth, *Schistosoma mansoni, J. Exp. Med.* **173**:159.
4. Sher, A., R. L. Coffman, S. Hieny, and A. W. Cheever, 1990, Ablation of eosinophil and IgE responses with anti-IL-5 or anti-IL-4 antibodies fails to affect immunity against Schistosoma mansoni in the mouse, *J. Immunol.* **145**:3911.
5. Mosmann, T. R., and R. L. Coffman, 1989, Heterogeneity of cytokine secretion patterns and functions of helper T cells, *Adv. Immunol.* **46**:111.
6. Wynn, T. A., I. Eltoum, A. W. Cheever, F. A. Lewis, W. C. Gause, and A. Sher, 1993, Analysis of cytokine mRNA expression during primary granuloma formation induced by eggs of *Schistosoma mansoni, J. Immunol.* **151**:1430.
7. Vella, A. T., and E. J. Pearce, 1992, CD4+ Th2 response induced by Schistosoma mansoni eggs develops rapidly, through an early, transient, Th0-like stage, *J. Immunol.* **148**:2283.
8. Sabin, E. A., and E. J. Pearce, 1995, Early IL-4 production by non-CD4+ cells at the site of antigen deposition predicts the development of a T helper 2 cell response to *Schistosoma mansoni* eggs, *J. Immunol.* **155**:4844.
9. Okano, M., K. Nishizaki, M. Abe, M. M. Wang, T. Yoshino, A. R. Satoskar, Y. Masuda, and D. A. Harn, Jr, 1999, Strain-dependent induction of allergic rhinitis without adjuvant in mice, *Allergy* **54**:593.
10. Lukacs, N. W., R. M. Strieter, S. W. Chensue, and S. L. Kunkel, 1994, Interleukin-4-dependent pulmonary eosinophil infiltration in a murine model of asthma, *Am. J. Respir. Cell. Mol. Biol.* **10**:526.
11. Okano, M., A. R. Satoskar, K. Nishizaki, M. Abe, and D. A. Harn, Jr., 1999, Induction of Th2 responses and IgE is largely due to carbohydrates functioning as adjuvants on Schistosoma mansoni egg antigens, *J. Immunol.* **163**:6712.
12. Brunet, L. R., F. D. Finkelman, A. W. Cheever, M. A. Kopf, and E. J. Pearce, 1997, IL-4 protects against TNF-alpha-mediated cachexia and death during acute schistosomiasis, *J. Immunol.* **159**:777.
13. Fallon, P. G., E. J. Richardson, G. J. McKenzie, and A. N. McKenzie, 2000, Schistosome infection of transgenic mice defines distinct and contrasting pathogenic roles for IL-4 and IL-13: IL-13 is a profibrotic agent. *J. Immunol.* **164**:2585.
14. Hoffmann, K. F., A. W. Cheever, and T. A. Wynn, 2000, IL-10 and the dangers of immune polarization: excessive type 1 and type 2 cytokine responses induce distinct forms of lethal immunopathology in murine schistosomiasis, *J. Immunol.* **164**:6406.
15. Hoffmann, K. F., T. A. Wynn, and D. W. Dunne, 2002, Cytokine-mediated host responses during schistosome infections; walking the fine line between immunological control and immunopathology, *Adv. Parasitol.* **52**:265.
16. Stadecker, M. J., H. J. Hernandez, and H. Asahi, 2001, The identification and characterization of new immunogenic egg components: implications for evaluation and control of the immunopathogenic T cell response in schistosomiasis, *Mem. Inst. Oswaldo Cruz* **96 Suppl**:29.

17. Faveeuw, C., T. Mallevaey, K. Paschinger, I. B. Wilson, J. Fontaine, R. Mollicone, R. Oriol, F. Altmann, P. Lerouge, M. Capron, and F. Trottein, 2003, Schistosome N-glycans containing core alpha 3-fucose and core beta 2-xylose epitopes are strong inducers of Th2 responses in mice, *Eur. J. Immunol.* **33**:1271.

18. Van der Kleij, D., A. Van Remoortere, J. H. Schuitemaker, M. L. Kapsenberg, A. M. Deelder, A. G. Tielens, C. H. Hokke, and M. Yazdanbakhsh, 2002, Triggering of innate immune responses by schistosome egg glycolipids and their carbohydrate epitope GalNAc beta 1-4(Fuc alpha 1-2Fuc alpha 1-3)GlcNAc, *J. Infect. Dis.* **185**:531.

19. Okano, M., A. R. Satoskar, K. Nishizaki, and D. A. Harn, Jr., 2001, Lacto-N-fucopentaose III found on Schistosoma mansoni egg antigens functions as adjuvant for proteins by inducing Th2-type response, *J. Immunol.* **167**:442.

20. Murphy, K. M., and S. L. Reiner., 2002, The lineage decisions of helper T cells, *Nat. Rev. Immunol.* **2**:933.

21. Trinchieri, G., 2003, Interleukin-12 and the regulation of innate resistance and adaptive immunity, *Nat. Rev. Immunol.* **3**:133.

22. Robinson, D. S., and A. O'Garra, 2002, Further checkpoints in Th1 development, *Immunity* **16**:755.

23. Pearce, E. J., A. Cheever, S. Leonard, M. Covalesky, R. Fernandez-Botran, G. Kohler, and M. Kopf, 1996, *Schistosoma mansoni* in IL-4-deficient mice, *Int. Immunol.* **8**:435.

24. Patton, E. A., L. R. Brunet, A. C. La Flamme, J. Pedras-Vasconcelos, M. Kopf, and E. J. Pearce, 2001, Severe schistosomiasis in the absence of interleukin-4 (IL-4) is IL-12 independent, *Infect. Immun.* **69**:589.

25. Sabin, E. A., M. A. Kopf, and E. J. Pearce, 1996, *Schistosoma mansoni* egg-induced early IL-4 production is dependent upon IL-5 and eosinophils, *J. Exp. Med.* **184**:1871.

26. Haisch, K., G. Schramm, F. H. Falcone, C. Alexander, M. Schlaak, and H. Haas, 2001, A glycoprotein from *Schistosoma mansoni* eggs binds non-antigen-specific immunoglobulin E and releases interleukin-4 from human basophils, *Parasite Immunol.* **23**:427.

27. Schramm, G., F. H. Falcone, A. Gronow, K. Haisch, U. Mamat, M. J. Doenhoff, G. Oliveira, J. Galle, C. A. Dahinden, and H. Haas, 2003, Molecular characterization of an interleukin-4-inducing factor from *Schistosoma mansoni* eggs, *J. Biol. Chem.* **278**:18384.

28. Brunet, L. R., E. A. Sabin, A. W. Cheever, M. A. Kopf, and E. J. Pearce, 1999, Interleukin 5 (IL-5) is not required for expression of a Th2 response or host resistance mechanisms during murine schistosomiasis mansoni but does play a role in development of IL-4-producing non-T, non-B cells, *Infect. Immun.* **67**:3014.

29. Brown, D. R., D. J. Fowell, D. B. Corry, T. A. Wynn, N. H. Moskowitz, A. W. Cheever, R. M. Locksley, and S. L. Reiner, 1996, Beta 2-microglobulin-dependent NK1.1+ T cells are not essential for T helper cell 2 immune responses, *J. Exp. Med.* **184**:1295.

30. Schmitz, J., A. Thiel, R. Kuhn, K. Rajewsky, W. Muller, M. Assenmacher, and A. Radbruch, 1994, Induction of interleukin 4 (IL-4) expression in T helper (Th) cells is not dependent on IL-4 from non-Th cells, *J. Exp. Med.* **179**:1349.

31. Mellman, I., and R. M. Steinman, 2001, Dendritic cells: specialized and regulated antigen processing machines, *Cell* **106**:255.

32. Shortman, K., and Y. J. Liu, 2002, Mouse and human dendritic cell subtypes. *Nat. Rev. Immunol.* **2**:151.

33. MacDonald, A. S., A. D. Straw, B. Bauman, and E. J. Pearce, 2001, CD8-dendritic cell activation status plays an integral role in influencing Th2 response development, *J. Immunol.* **167**:1982.

34. de Jong, E. C., P. L. Vieira, P. Kalinski, J. H. Schuitemaker, Y. Tanaka, E. A. Wierenga, M. Yazdanbakhsh, and M. L. Kapsenberg, 2002, Microbial compounds selectively induce

Th1 cell-promoting or Th2 cell-promoting dendritic cells in vitro with diverse th cell-polarizing signals, *J. Immunol.* **168**:1704.

35. Faveeuw, C., V. Angeli, J. Fontaine, C. Maliszewski, A. Capron, L. Van Kaer, M. Moser, M. Capron, and F. Trottein, 2002, Antigen presentation by CD1d contributes to the amplification of Th2 responses to *Schistosoma mansoni* glycoconjugates in mice, *J. Immunol.* **169**:906.

36. Oswald, I. P., P. Caspar, D. Jankovic, T. A. Wynn, E. J. Pearce, and A. Sher, 1994, IL-12 inhibits Th2 cytokine responses induced by eggs of *Schistosoma mansoni*, *J. Immunol.* **153**:1707.

37. Zaccone, P., Z. Fehervari, F. M. Jones, S. Sidobre, M. Kronenberg, D. W. Dunne, and A. Cooke, 2003, *Schistosoma mansoni* antigens modulate the activity of the innate immune response and prevent onset of type 1 diabetes, *Eur. J. Immunol.* **33**:1439.

38. Scanga, C. A., J. Aliberti, D. Jankovic, F. Tilloy, S. Bennouna, E. Y. Denkers, R. Medzhitov, and A. Sher, 2002, Cutting edge: MyD88 is required for resistance to *Toxoplasma gondii* infection and regulates parasite-induced IL-12 production by dendritic cells, *J. Immunol.* **168**:5997.

39. Aliberti, J., J. G. Valenzuela, V. B. Carruthers, S. Hieny, J. Andersen, H. Charest, C. Reis e Sousa, A. Fairlamb, J. M. Ribeiro, and A. Sher, 2003, Molecular mimicry of a CCR5 binding-domain in the microbial activation of dendritic cells, *Nat. Immunol.* **4**:485.

40. Jankovic, D., M. C. Kullberg, S. Hieny, P. Caspar, C. M. Collazo, and A. Sher, 2002, In the absence of IL-12, CD4(+) T cell responses to intracellular pathogens fail to default to a Th2 pattern and are host protective in an IL-10(-/-) setting, *Immunity* **16**:429.

41. Blander, J. M., D. B. Sant'Angelo, K. Bottomly, and C. A. Janeway, Jr., 2000, Alteration at a single amino acid residue in the T cell receptor alpha chain complementarity determining region 2 changes the differentiation of naive CD4 T cells in response to antigen from T helper cell type 1 (Th1) to Th2, *J. Exp. Med.* **191**:2065.

42. Leitenberg, D., F. Balamuth, and K. Bottomly, 2001, Changes in the T cell receptor macromolecular signaling complex and membrane microdomains during T cell development and activation, *Semin. Immunol.* **13**:129.

43. MacDonald, A. S., and E. J. Pearce, 2002, Cutting edge: polarized Th cell response induction by transferred antigen-pulsed dendritic cells is dependent on IL-4 or IL-12 production by recipient cells, *J. Immunol.* **168**:3127.

44. Steinman, R. M., D. Hawiger, K. Liu, L. Bonifaz, D. Bonnyay, K. Mahnke, T. Iyoda, J. Ravetch, M. Dhodapkar, K. Inaba, and M. Nussenzweig, 2003, Dendritic cell function in vivo during the steady state: a role in peripheral tolerance, *Ann. N Y Acad. Sci.* **987**:15.

45. MacDonald, A. S., A. D. Straw, N. M. Dalton, and E. J. Pearce, 2002, Cutting edge: Th2 response induction by dendritic cells: a role for CD40, *J. Immunol.* **168**:537.

46. MacDonald, A. S., E. A. Patton, A. C. La Flamme, M. I. Araujo, C. R. Huxtable, B. Bauman, and E. J. Pearce, 2002, Impaired Th2 development and increased mortality during *Schistosoma mansoni* infection in the absence of CD40/CD154 interaction, *J. Immunol.* **168**:4643.

47. Straw, A. D., A. S. MacDonald, E. Y. Denkers, and E. J. Pearce, 2003, CD154 plays a central role in regulating dendritic cell activation during infections that induce Th1 or Th2 responses, *J. Immunol.* **170:727-734**.

48. Sher, A., D. Fiorentino, P. Caspar, E. Pearce, and T. Mosmann, 1991, Production of IL-10 by CD4+ T lymphocytes correlates with down-regulation of Th1 cytokine synthesis in helminth infection, *J. Immunol.* **147**:2713.

49. Shevach, E. M, 2002, CD4+ CD25+ suppressor T cells: more questions than answers, *Nat. Rev. Immunol.* **2**:389.

50. van der Kleij, D., E. Latz, J. F. Brouwers, Y. C. Kruize, M. Schmitz, E. A. Kurt-Jones, T. Espevik, E. C. de Jong, M. L. Kapsenberg, D. T. Golenbock, A. G. Tielens, and M. Yazdanbakhsh, 2002, A novel host-parasite lipid cross-talk. Schistosomal lyso-phosphatidylserine activates toll-like receptor 2 and affects immune polarization, *J. Biol. Chem.* **277**:48122.

51. Van Die, I., S. J. Van Vliet, A. K. Nyame, R. D. Cummings, C. M. Bank, B. Appelmelk, T. B. Geijtenbeek, and Y. Van Kooyk, 2003, The dendritic cell-specific C-type lectin DC-SIGN is a receptor for *Schistosoma mansoni* egg antigens and recognizes the glycan antigen Lewis x, *Glycobiology* **13**:471.

52. Steinman, R. M., 1996, Dendritic cells and immune-based therapies, *Exp. Hematol.* **24**:859.

53. Tailleux, L., O. Schwartz, J. L. Herrmann, E. Pivert, M. Jackson, A. Amara, L. Legres, D. Dreher, L. P. Nicod, J. C. Gluckman, P. H. Lagrange, B. Gicquel, and O. Neyrolles, 2003, DC-SIGN is the major Mycobacterium tuberculosis receptor on human dendritic cells, *J. Exp. Med.* **197**:121.

Chapter 9

IMMUNOPATHOLOGY IN EXPERIMENTAL SCHISTOSOMIASIS

Thomas A. Wynn, Allen W. Cheever, Mallika Kaviratne, Robert W. Thompson, Margaret M. Mentink-Kane, and Matthias Hesse
Laboratory of Parasitic Diseases, National Institute of Allergy and Infectious Diseases, National Institutes of Health, 50 South Drive, Rm.6154, MSC 8003, Bethesda, MD 20892

Key words: fibrosis, granuloma, Th1/Th2, liver, collagen, suppression, dendritic cell, regulatory T cells, apoptosis, vaccine

1.   PATHOGENESIS

In *Schistosomiasis mansoni*, parasites migrate to the mesenteric veins where they begin laying hundreds of eggs per day approximately 4 to 5 weeks post-infection. Some eggs are trapped in the microvasculature of the liver where they induce a vigorous granulomatous response. Subsequently, in severe cases fibrosis and portal hypertension develop. These are the primary causes of morbidity in infected individuals and may eventually be lethal due to variceal bleeding. Consequently, much of the symptomatology of schistosomiasis is attributed to the egg-induced granulomatous response and the associated pathology.   Although the magnitude and location of pathology differs with other schistosome species, the pathogenic mechanisms are likely similar. Therefore, given the wealth of information on the immunopathology of *S. mansoni* infection, we have restricted our discussion to recent progress in this area.

As mentioned above hepatic granulomas are pathogenic because they precipitate fibrosis, which obstructs blood flow, increases portal blood pressure, and ultimately promotes development of portal-systemic venous shunts. CD4$^+$ Th cells are essential for granuloma formation, while all other lymphocytes examined so far (including B cells, CD8$^+$ T cells, NK T cells, and $\gamma\delta$ T cells) do not appear to be involved, at least in the early initiation phases of granuloma development (1). Studies aimed at dissecting the respective roles of type-1- (IFN-γ, IL-2, TNF-α) and type-2 CD4$^+$ T cell-associated cytokines (IL-4, IL-5, IL-6, IL-9, IL-13) in granuloma formation showed that the granulomatous response evolves from an early type-1 to a sustained and dominant type-2 cytokine response (2). The importance of CD4$^+$ Th2 cells to the pathogenesis of schistosomiasis was shown in experiments in which mice vaccinated with egg antigen extracts plus IL-12 to induce an egg-specific CD4$^+$ Th1 response upon subsequent infection, developed smaller granulomas and less severe fibrosis than did non-vaccinated Th2-polarized controls (3-6). Decreased fibrosis was associated with a diminished type-2 response and increased type-1 cytokine production.

Although granulomas are detrimental because they eventually "scar" the liver, it is clear that the egg-induced lesions also serve an important host-protective function, particularly in *S. mansoni* infections. During infection, antigens secreted by schistosome eggs provide a continuous and potent stimulus for the immune response. If these antigens are not sequestered or neutralized effectively, they may trigger a persistent and tissue damaging host immune response. In support of this conclusion, CD4$^+$ T-cell-deprived, egg-tolerized, and some IL-4$^{-/-}$ and IL-10$^{-/-}$ mice die earlier than comparably infected, immunologically intact control mice because they are unable to satisfactorily mount a normal type-2-dominant response (7). Widespread microvesicular hepatic damage induced by toxic egg products contributed to the death of the infected mice, particularly in the case of nude and SCID mice (8). Presumably, the detrimental effect of chronic type-2 cytokine expression (e.g. fibrosis and portal hypertension) represents a compromise solution for the parasite and host as the parasite can persist only when the host survives. Because of the central role played by type-2 cytokines, much work on the pathogenesis of schistosomiasis is focused on the mechanisms that initiate, maintain, and suppress type-2 immunity.

2.   THE GENERATION OF TYPE-2 RESPONSES

Type-2-defective Stat6, IL-4R, and IL-4/IL-13-deficient mice are all severely impaired in granuloma formation and fibrosis while type-1-defective Stat1- (unpublished observation), Stat4-, IFN-γ-, and IL-12p40-deficient mice show only minor changes in pathology (9-11), confirming the critical role of type-2 cytokines in disease progression. Surprisingly, while the mechanisms that initiate type-1 immunity have been described in detail in a variety of models (12), the mechanisms that promote type-2 responses are much less clear. Many studies suggested that accessory cells such as mast cells, basophils, Kupffer cells, eosinophils, or non-B, non-T cells expressing Fc epsilon receptors might be important by providing an early source of IL-4 (13), however no study has demonstrated a convincing role for one particular cell, suggesting multiple pathways may be involved in the initiation of type-2 immunity, including an autocrine role for T cells (14). It is clear however, that IL-4, Stat6, and the IL-4 receptor are required to stabilize and amplify the Th2 response once initiated (15). Recent studies have also documented an important role for dendritic cells (16), with IL-10 likely serving as a key regulatory factor (17, 18). Thus, multiple mechanisms appear to be involved in the initiation, amplification, and stabilization of Th2 responses. In the coming years an important and growing area of research will be to better understand how these various processes are triggered initially, connected mechanistically, and regulated over time in schistosomiasis. The effector functions of the entire family of type-2 cytokines also need further dissection. Chapter 8 provides a detailed discussion on initiation of type-2 immunity in schistosomiasis.

3. ACUTE TOXEMIC SCHISTOSOMIASIS (KATAYAMA FEVER)

Many individuals, usually from outside endemic areas and therefore naive to schistosome antigens, develop acute disease 3-6 weeks after their first exposure to schistosome infection (19). Fever, malaise (including extreme prostration) anorexia and marked eosinophilia are common and may occur before egg laying begins, becoming worse with the onset of oviposition. The severity of acute toxemic schistosomiasis is related to the intensity of infection, estimated from the number of eggs in the feces, but lightly infected patients are sometimes very ill (19). Symptoms and signs, probably caused by circulating immune complexes (20) and TNF-α (21), improve while the schistosome infection continues unchecked. Acute toxemic infections are associated with florid, large granulomas around eggs deposited in the tissues.

Acute toxemic disease in humans is presumably unrelated to toxins secreted by the eggs of *S. mansoni* described by Dunne et al. (22) since disease often precedes egg deposition and toxemic disease is frequent in *S. japonicum* infections of humans but no analogous egg toxin is found in *S. japonicum* eggs (23). Toxemic disease is also distinguished from cytokine shock (24), frequently causing cachexia and death that are uncommon in acute toxemic schistosomiasis. Additionally, tissue and blood eosinophilia are increased in acute toxemic schistosomiasis of humans but decreased in mice with cytokine shock. Signs of cytokine shock are also absent in immunodeficient mice infected with schistosome species other than *S.*

mansoni (24). In contrast, symptoms in acute toxemic disease are related to levels of circulating immune complexes, occur in response to a variety of schistosome species, and are compatible with immune complex disease. De Jesus et al. recently examined cases of acute toxemic schistosomiasis in Brazilian patients and attributed symptoms to proinflammatory cytokines, immune complexes, and low Th2 responses to infection (21).

Most laboratory animals develop ruffled fur, decreased activity and marked eosinophilia during the first 2 months of infection. Unless the infection intensity is overwhelming, these symptoms and signs subside as the infection becomes chronic. Primates seem the best model of acute disease and Kanamura et al. (25) reported fever and elevated levels of TNF-α, IL-1 and IL-6 in *S. mansoni*-infected baboons and Sadun et al. noted transient fever and illness in *S. mansoni* infected chimpanzees (26). Although immunologic downregulation of host responses have been intensively examined in the mouse (7), no detailed comparison has been made with human acute toxemic schistosomiasis and its downregulation.

4.   CYTOKINE SHOCK

As mentioned previously, at the onset of granulomatous inflammation, the dominant CD4$^+$ T cell response changes from a type-1 response of short duration to a sustained type-2 response. The development of Th2 cells *in vitro* is highly dependent on IL-4; therefore, it was originally predicted that IL-4$^{-/-}$ mice might develop less severe disease. However, C57BL/6 IL-4$^{-/-}$ mice suffer from severe acute disease characterized by cachexia and significant mortality, unlike WT mice that develop chronic infections. The primary cause of morbidity in the infected IL-4$^{-/-}$ animals was attributed to hemorrhagic lesions in the mucosa of the small intestine (27), which gradually increases host exposure to bacteria and bacterial toxins that reside in the gut (28). In addition, the decreased type-2 and enhanced type-1 response results in increased proinflammatory cytokine expression which is thought to contribute to the weight loss and death of the infected IL-4$^{-/-}$ mice (27, 29). Thus, the mortality is reminiscent of cytokine-induced shock seen during sepsis. In support of this hypothesis, morbidity is partially ameliorated when infected IL-4$^{-/-}$ animals are treated with a neutralizing antibody against TNF-α (27). Surprisingly however, IFN-γ and IL-12 appear to play little or no role since C57BL/6 IL-4/IL-12p35$^{-/-}$ and anti-IFN-γ-treated IL-4$^{-/-}$ animals developed the same disease as IL-4$^{-/-}$ mice (30, 31). It is important to note, however, that not every study examining IL-4$^{-/-}$ mice reported increased mortality, which may be due to the parasite burden used or genetic background of the IL-4$^{-/-}$ host (17, 32, 33).

Increased IFN-γ cytokine production was also observed in IL-10$^{-/-}$ mice, suggesting that IL-10 and IL-4 co-operate to limit morbidity in the early stages of schistosomiasis by inducing and stabilizing the type-2 response (17, 34). Indeed, marked increases in IFN-γ, TNF-α, and nitric oxide (NO) were detected in infected IL-10$^{-/-}$ animals and the mice developed significant morbidity and mortality. Strikingly however, mice deficient in both IL-4 and IL-10 were much more susceptible and uniformly died between weeks seven and nine post-infection. The double cytokine-deficient animals also developed the strongest and most highly polarized type-1 response; in addition, elevated serum transaminase levels suggested mortality was attributable, in part, to acute hepatotoxicity. cDNA microarray experiments on granulomatous tissues suggested that neutrophil-induced damage and apoptosis contribute to the severe pathology of infected type-1 polarized mice (35). Other studies have suggested that IL-4, and perhaps IL-10, protect the host by downregulating the production of reactive oxygen and nitrogen intermediates that damage the liver (29). Consistent with this conclusion, disease severity in IL-4$^{-/-}$ mice correlated with the level of nitric oxide (NO) produced by LPS-activated spleen cells (36) and infected mice treated with uric acid, a scavenger of ONOO(-), lived longer and developed less pathology (29). Collectively, these observations demonstrate that IL-10 and IL-4 are both required to prevent type-1 responses from becoming pathological during acute *S. mansoni* infection. Although the exact cause of tissue pathology and death in these immunodeficient mice likely varies, the mechanisms involved have been elegantly worked out. Nevertheless, cytokine shock seems to us to not be related to acute or chronic schistosomiasis in humans and the absence of similar syndromes in immunodeficient mice infected with other schistosome species supports this conclusion (24). Conceivably, cytokine shock might develop in patients with immunodepression such as that associated with AIDS but the intensity of infection in most humans is miniscule compared to infection intensity in a mouse infected with a single worm pair (37). A recent study showed that patients with schistosomiasis/HIV-1 coinfections produced significantly less IL-4 and IL-10 and had increased Th1:Th2 cytokine ratios than did HIV-1-negative individuals (38). Nevertheless, the very low levels of infection in humans compared with those in mice would suggest that egg-mediated cytokine shock should seldom, if ever, occur even in coinfected patients.

5.   TYPE-2 CYOTKINES INDUCE FIBROTIC PATHOLOGY IN CHRONIC SCHISTOSOMIASIS

Severe chronic schistosomiasis mansoni in man is characterized by Symmers' pipestem periportal fibrosis (39) and type-2 cytokines are central to the development of hepatic fibrosis in mice (7). Limited data from mice and humans suggest type-1-associated cytokines are involved in the pathogenesis of Symmers' fibrosis, although TNF-α may play a role (40-42). High plasma levels of sTNFR-I, sTNFR-II, and ICAM-1 have also been reported in patients with hepatosplenomegaly. Thus, while the type-2 response plays a protective role in the initial stages of infection, persistent expression of these mediators may cause severe fibrosis and portal hypertension in some chronically infected individuals. The type-2 response therefore represents a "double-edged sword", by exhibiting protective and host-damaging activities during acute and chronic infection, respectively.

Microscopic, biochemical, and molecular techniques all indicated that IL-13, but not IL-4, plays the major role in the development of egg-induced liver fibrosis (43).   Strikingly, mortality is delayed in infected IL-13$^{-/-}$ animals, further highlighting the major contribution of IL-13 to the pathogenesis of chronic schistosomiasis (28).   *In vitro* studies also demonstrated the ability of IL-13 to stimulate collagen production in fibroblasts (43). Thus, the effects of IL-13 on fibrosis may be direct, and not dependent upon other pro-fibrotic mediators. Studies conducted on normal human skin and keloid fibroblasts also suggested a direct role for IL-13 (44). Nevertheless, additional studies are needed to determine whether other mediators, such as TGF-beta, are involved (45).

Studies   conducted   in   IL-10$^{-/-}$/IL-12$^{-/-}$ and   IL-10$^{-/-}$/IFN-γ$^{-/-}$ mice underscored the central role of IL-13 in fibrogenesis.   These animals developed a highly exaggerated IL-13 response following infection with *S. mansoni* (17, 46) and hepatic fibrosis increased markedly. The double cytokine-deficient mice also displayed significant morbidity and mortality during chronic infection. Blood was frequently found in the intestine at autopsy, which suggested that portal hypertension and intestinal bleeding contributed to their death. These observations suggest that IL-10, IL-12, and IFN-γ are all required to prevent type-2 responses from overshooting and becoming pathological in chronic infections (17, 47). Thus, in murine schistosomiasis, distinct but equally detrimental forms of lethal tissue pathology develop when the immune response is biased toward an extreme type-1 or type-2 phenotype (35). Therefore, co-dominant type-1/type-2 responses or weakly polarized type-1 responses that are controlled by IL-10 or regulatory T cells might offer the best protection from egg-induced liver pathology during both acute and chronic schistosomiasis (Figure 1).

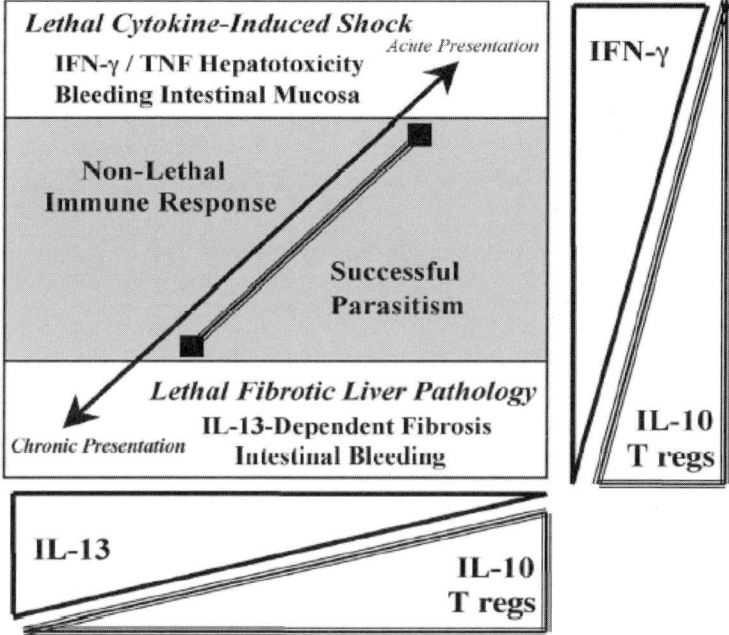

Figure 9-1. Pathogenesis of Experimental Schistosomiasis. Distinct forms of lethal tissue pathology arise when *S. mansoni* infected mice develop chronic or highly polarized type-1 or type-2 immune responses. The position on the solid line symbolizes the host immune response at any given time following infection. The triangles on the right and bottom represent the magnitude of the IFN-g and IL-13 response, respectively. The dashed triangles denote the IL-10 or regulatory T cell response and the dashed line and box insert represent the "optimal zone" where morbidity or mortality is minimized. The infected host remains in the optimal zone as long as "immune-downmodulation" is preserved, perhaps through multiple overlapping mechanisms.

6. ENDOGENOUS MECHANISMS THAT REGULATE THE PROGRESSION OF SCHISTOSOMIASIS

The robust granulomatous response that peaks between 8 and 9 wk is gradually downmodulated as the infection progresses to the chronic stage. It is widely accepted that the failure of some individuals to effectively induce this downmodulated state might explain the development of hepatosplenic disease. Although several mechanisms to explain the process of "immune modulation" have been described, it has been difficult to unite all of the published findings into one general mechanism or pathway.

6.1   Anergy, cross-reactive idiotypes, regulatory T cells, and Interleukin-10

In one model, it was hypothesized that immunomodulation might represent a state of anergy in the responding Th1 cell (48). Macrophages derived from hepatic granulomas induced "anergy" in cultured Th1-cell clones and down-regulated pulmonary granuloma formation *in vivo* (49), with IL-10 serving as the putative mediator (50). Additional support for this hypothesis came from early studies that showed anti-IL-10 Ab could restore IFN-γ production in splenocyte cultures from infected mice (51). Nevertheless, macrophage-induced anergy in the Th1 cell population cannot fully explain immune downmodulation since much of the pathology is driven by type-2 cytokines (45), yet $CD4^+$ Th2 cells were initially thought to be refractory to the anergy-inducing activities of macrophages (52).

Other studies confirmed the critical role of IL-10 in the progression of schistosomiasis (17, 34, 53-55), with IL-10 exhibiting potent suppressive activity on both type-1 and type-2 responses (17, 46). In one study, schistosome-specific cross-reactive idiotypic antibodies (Id) produced during schistosome infection were shown to modulate egg-induced pathology and this correlated with the production of high levels of IL-10. In this model, approximately 20% of infected male CBA/J mice develop a hypersplenomegaly syndrome (HSS) while the rest present with moderate splenomegaly syndrome (MSS), thus mirroring the severe hepatosplenic and moderate intestinal forms of the disease in humans (53, 56). Interestingly, protective Id preparations prepared from chronic MSS mice upregulated IFN-γ (57). Spleen cells from MSS mice also produced significantly more IL-10 than did mice with HSS (53). This association of low levels of antigen-induced IL-10 and IFN-γ with severe pathology (HSS) is thus consistent with the fibrotic pathology observed in IL-10/IFN-γ and IL-10/IL-12 double-deficient mice (17, 46).   Moreover, a recent study comparing egg-induced pathology in mice that develop either modest or severe fibrosis showed that severe disease correlates with a high IL-13 and low IFN-γ/IL-10 response (47). Importantly, perinatal exposure to MSS Ids promoted long-term survival in mice subsequently infected with *S. mansoni*, and this was associated with increased IFN-γ production, further illustrating the potential host protective function of IFN-γ in chronic schistosomiasis (58).   The critical events that trigger protective Ids in some schistosome infections, however, remain unclear (59). Nevertheless, they illustrate a promising and powerful strategy that might be exploited to slow disease progression.

Recent studies in related asthma/allergy models have suggested that regulatory T cells producing IL-10 and/or TGF-beta might also play an essential inhibitory role during type-2 dominant immune responses. In one study, dendritic cells exposed to a defined respiratory antigen were shown to transiently produce IL-10, which in turn stimulated the development of IL-10-producing CD4[+] T regulatory cells (Tr1). Thus, induction of tolerance by IL-10 may provide an effective means to suppress a variety of type-2 cytokine-dominant diseases. Nevertheless, it is important to note that other studies reported increased airway hyperreactivity, eosinophilic inflammation, and Th2 effector function in response to IL-10 (45). Similar findings were also reported in the *Trichuris muris* model, in which IL-10 promoted IL-4/IL-13-mediated parasite expulsion by suppressing the counter-regulatory type-1 cytokine response. Therefore, additional investigations are needed before induction of IL-10 or regulatory T cells can be advocated as a strategy to treat or prevent IL-4/IL-13-driven disorders. The important contribution of IL-10-producing dendritic cells, macrophages, and regulatory T cells in the induction, maintenance, and resolution of schistosomiasis was recently shown (18).

6.2   B cells, antibodies, and apoptosis

In addition to studies on protective idiotypes (57), early B cell depletion studies were the first to suggest a possible role for B cells in immune down-modulation (60). Recent studies with B cell-deficient mice confirmed these observations (61, 62). A more recent study reported decreased apoptosis of CD4[+] T-cells in infected B-cell deficient mice, providing a possible explanation for their exacerbated granulomatous response. These authors showed that Fas ligand (FasL) expressing B cells were capable of stimulating significant CD4[+] T-cell apoptosis, with large numbers of FasL+ B cells being generated following infection (63, 64). Interestingly, FasL expression on B cells was inhibited by anti-IL-10 and anti-IL-4 mAb and upregulated by rIL-4 and rIL-10 (64), suggesting a possible link between the type-2 cytokine response and activation-induced cell death (AICD). These studies suggested that B cells regulate the pathogenesis of schistosomiasis by influencing the magnitude of the antigen-specific CD4[+] T cell population. A recent study also reported no impairment in CD4[+] T cell apoptosis in infected Fas- and FasL-deficient mice (65). Instead, decreased production of the T cell growth factor IL-2 was proposed as a possible explanation for increased apoptosis. These exciting findings suggest more detailed study is needed on the role of B cells and apoptosis in chronic schistosomiasis.

6.3   The interleukin-13 decoy receptor

Importantly, recent studies conducted with knockout mice confirmed that the IL-13Rα2 is a functional decoy receptor for the Th2 response (66). The regulation and function of the IL-13Rα2 was also examined in murine schistosomiasis (67). Shortly after the onset of egg-laying a marked upregulation of 13Rα2 expression is observed. Receptor expression is highly dependent on IL-10, IL-13, and Stat6 and inhibited by the Th1-inducing cytokine IL-12, thus revealing a strong correlation with the egg-induced type-2 response. Strikingly, when IL-13Rα2-deficient mice were infected with *S. mansoni*, they showed a marked exacerbation in IL-13-dependent fibrosis and completely failed to downregulate granuloma formation during chronic infection (67, 68). These studies were important because they revealed a novel endogenous mechanism that limits Th2-dependent fibrotic pathology and identified a novel pathway of immune suppression. Given these important findings, additional studies investigating the role of the decoy receptor in human schistosomiasis are warranted. Finally, given that several distinct mechanisms may explain the process of immune down-modulation, future efforts should focus on elucidating how these various pathways are connected mechanistically.

7.   IMMUNE DEVIATION, ANTI-PATHOLOGY VACCINES, AND ALTERNATIVELY-ACTIVATED MACROPHAGES

Over the past several years we have used immune deviation as a tool to study the mechanisms of tissue remodeling and fibrosis. Several mediators and cell types that characterize type-1 cytokine responses, including TNF-α, IL-12, and NO, inhibit hepatic fibrosis (5, 69). Because macrophages and dendritic cells are key sources of these mediators, recent efforts have focused on elucidating their respective roles in pathogenesis (69, 70). Unexpectedly, the beneficial effects of type-2 to type-1 immune-deviation were completely lost in inducible nitric oxide synthase-2 (NOS-2)-deficient mice, despite the development of a type-1 polarized response. In fact, egg-induced liver pathology worsened in the immune-deviated knockout animals (69). These studies were the first to suggest that the phenotype of the macrophage/dendritic cell might be more critical to the regulation of egg-induced pathology than the antigen specific CD4+ T helper cell.

Macrophages are a prominent constituent of granulomas. Interestingly, recent studies showed that macrophages express two cytokine inducible enzymes, NOS-2 and arginase, which share L-arginine as a substrate. In studies by Modolell and colleagues, different combinations of type-1 and type-2 cytokines induced NOS-2 and arginase-1 (Arg-1) expression (71). The Th1-associated cytokines IFN-γ and TNF-α induced NOS-2 in macrophages designated "classically-activated", whereas the type-2 cytokines induced Arg-1 in cells termed "alternatively-activated" (72).

Strikingly, recent studies confirmed the strict requirement for type 2 cytokines in the inducible expression and activation of Arg-1 *in vivo*, and in agreement with early *in vitro* data (71), IL-13 was identified as a major arginase inducer in mice exposed to schistosome eggs (70). Type-2 cytokines also stimulated proline production in macrophages by an arginase dependent mechanism (70). While it is not yet known which cells in the granuloma, besides macrophages, produce Arg-1 and proline, high arginase activity correlated consistently with the formation of large granulomas (70). In contrast, NOS-2 was induced in mice sensitized with eggs and IL-12, and these animals developed smaller granulomas with less fibrosis. These data confirmed previous *in vitro* findings and served as a first example of the differential regulation of NOS-2/Arg-1 by type-1/type-2 cytokines in granulomatous disease (70). They also explained how CD4+ Th2 cells and alternatively activated macrophages might co-operate to control the overall magnitude of fibrosis. Further study of this fibrogenic pathway is needed to see whether it can be exploited in vaccine design.

8.   CONCLUDING REMARKS

Data generated from experimental models of schistosomiasis showed that hepatic fibrosis is tightly controlled by the pro-fibrotic and anti-fibrotic activities of IL-13 and IFN-γ, respectively. Interestingly, Dessein and colleagues recently investigated genetic control of severe schistosomiasis in Sudan. These investigators identified a major human locus closely linked to the IFN-gamma R1 gene, which encodes the receptor for IFN-γ (73). The authors' hypothesized polymorphisms within the IFN-gamma R1 gene might influence the development of hepatic disease. Polymorphisms within the IL-13 promoter (74) and IL-13Rα1 gene (75) were also recently described. Thus, IL-13-dependent signaling pathways might influence development of fibrosis, portal hypertension, and hepatosplenic disease in humans. Nevertheless, the regulation and function of IL-13 in human schistosomiasis remains mostly unexplored. Translating findings from the murine model of schistosomiasis to the human disease remains a high priority. This will accelerate progress towards our ultimate goal of designing a highly effective anti-infection and anti-pathology vaccine for this devastating disease.

ACKNOWLEDGEMENTS

We thank our past and present collaborators who participated in the work described. We are also grateful to Fred Lewis and his colleagues at the Biomedical Research Institute for their continuous support.

REFERENCES

1. Pearce, E. J., and A. S. MacDonald. 2002. The immunobiology of schistosomiasis. *Nat Rev Immunol 2:499.*
2. Grzych, J. M., E. J. Pearce, A. Cheever, Z. A. Caulada, P. Caspar, S. Heiny, F. Lewis, and A. Sher. 1991. Egg deposition is the major stimulus for the production of Th2 cytokines in murine schistosomiasis mansoni. *J.Immunol. 146:1322.*
3. Wynn, T. A., I. Eltoum, I. P. Oswald, A. W. Cheever, and A. Sher. 1994. Endogenous interleukin 12 (IL-12) regulates granuloma formation induced by eggs of Schistosoma mansoni and exogenous IL-12 both inhibits and prophylactically immunizes against egg pathology. *J Exp Med 179:1551.*
4. Wynn, T. A., A. W. Cheever, D. Jankovic, R. W. Poindexter, P. Caspar, F. A. Lewis, and A. Sher. 1995. An IL-12-based vaccination method for preventing fibrosis induced by schistosome infection. *Nature 376:594.*
5. Hoffmann, K. F., P. Caspar, A. W. Cheever, and T. A. Wynn. 1998. IFN-γ, IL-12, and TNF-α are required to maintain reduced liver pathology in mice vaccinated with Schistosoma mansoni eggs and IL-12. *J. Immunol. 161:4201.*
6. Boros, D. L., and J. R. Whitfield. 1999. Enhanced Th1 and dampened Th2 responses synergize to inhibit acute granulomatous and fibrotic responses in murine schistosomiasis mansoni. *Infect Immun 67:1187.*
7. Hoffmann, K. F., T. A. Wynn, and D. W. Dunne. 2002. Cytokine-mediated host responses during schistosome infections; walking the fine line between immunological control and immunopathology. *Adv Parasitol 52:265.*
8. Fallon, P. G. 2000. Immunopathology of schistosomiasis: a cautionary tale of mice and men. *Immunol Today 21:29.*
9. Wynn, T. A., D. Jankovic, S. Hieny, K. Zioncheck, P. Jardieu, A. W. Cheever, and A. Sher. 1995. IL-12 exacerbates rather than suppresses T helper 2-dependent pathology in the absence of endogenous IFN-gamma. *J Immunol 154:3999.*
10. Yap, G., A. Cheever, P. Caspar, D. Jankovic, and A. Sher. 1997. Unimpaired down-modulation of the hepatic granulomatous response in CD8 T-cell- and gamma interferon-deficient mice chronically infected with Schistosoma mansoni. *Infect Immun 65:2583.*
11. Kaplan, M. H., J. R. Whitfield, D. L. Boros, and M. J. Grusby. 1998. Th2 cells are required for the *Schistosoma mansoni* egg-induced granulomatous response. *J Immunol 160:1850.*
12. Szabo, S. J., B. M. Sullivan, S. L. Peng, and L. H. Glimcher. 2003. Molecular mechanisms regulating Th1 immune responses. *Annu Rev Immunol 21:713.*
13. Pearce, E. J., A. La Flamme, E. Sabin, and L. R. Brunet. 1998. The initiation and function of Th2 responses during infection with Schistosoma mansoni. *Adv Exp Med Biol 452:67.*
14. Coffman, R. L., and T. von der Weid. 1997. Multiple pathways for the initiation of T helper 2 (Th2) responses. *J Exp Med 185:373.*

15. Jankovic, D., M. C. Kullberg, N. Noben-Trauth, P. Caspar, W. E. Paul, and A. Sher. 2000. Single cell analysis reveals that IL-4 receptor/Stat6 signaling is not required for the in vivo or in vitro development of CD4+ lymphocytes with a Th2 cytokine profile. *J Immunol 164:3047.*

16. MacDonald, A. S., A. D. Straw, N. M. Dalton, and E. J. Pearce. 2002. Cutting edge: Th2 response induction by dendritic cells: a role for CD40. *J Immunol 168:537.*

17. Hoffmann, K. F., A. W. Cheever, and T. A. Wynn. 2000. IL-10 and the dangers of immune polarization: excessive type 1 and type 2 cytokine responses induce distinct forms of lethal immunopathology in murine schistosomiasis. *J. Immunol. 164:6406.*

18. Hesse, M., C. A. Piccirillo, Y. Belkaid, J. Prufer, M. Mentink-Kane, M. Leusink, A. W. Cheever, E. M. Shevach, and T. A. Wynn. 2004. The pathogenesis of schistosomiasis is controlled by cooperating IL-10-producing innate effector and regulatory T cells. *J. Immunol 172:00.*

19. Hiatt, R. A., Z. R. Sotomayor, G. Sanchez, M. Zambrana, and W. B. Knight. 1979. Factors in the pathogenesis of acute schistosomiasis mansoni. *J Infect Dis 139:659.*

20. Hiatt, R. A., E. A. Ottesen, Z. R. Sotomayor, and T. J. Lawley. 1980. Serial observations of circulating immune complexes in patients with acute schistosomiasis. *J Infect Dis 142:665.*

21. de Jesus, A. R., A. Silva, L. B. Santana, A. Magalhaes, A. A. de Jesus, R. P. de Almeida, M. A. Rego, M. N. Burattini, E. J. Pearce, and E. M. Carvalho. 2002. Clinical and immunologic evaluation of 31 patients with acute schistosomiasis mansoni. *J Infect Dis 185:98.*

22. Dunne, D. W., F. M. Jones, and M. J. Doenhoff. 1991. The purification, characterization, serological activity and hepatotoxic properties of two cationic glycoproteins (alpha 1 and omega 1) from Schistosoma mansoni eggs. *Parasitology 103 Pt 2:225.*

23. Cheever, A. W., J. A. Lenzi, H. L. Lenzi, and Z. A. Andrade. 2002. Experimental models of Schistosoma mansoni infection. *Mem Inst Oswaldo Cruz 97:917.*

24. Cheever, A. W., K. F. Hoffmann, and T. A. Wynn. 2000. Immunopathology of schistosomiasis mansoni in mice and men. *Immunol Today 21:465.*

25. Kanamura, H. Y., K. Hancock, V. Rodrigues, and R. T. Damian. 2002. Schistosoma mansoni heat shock protein 70 elicits an early humoral immune response in S. mansoni infected baboons. *Mem Inst Oswaldo Cruz 97:711.*

26. Sadun, E. H., F. von Lichtenberg, A. W. Cheever, and D. G. Erickson. 1970. Schistosomiasis mansoni in the chimpanzee. The natural history of chronic infections after single and multiple exposures. *Am J Trop Med Hyg 19:258.*

27. Brunet, L. R., F. D. Finkelman, A. W. Cheever, M. A. Kopf, and E. J. Pearce. 1997. IL-4 protects against TNF-alpha-mediated cachexia and death during acute schistosomiasis. *J Immunol 159:777.*

28. Fallon, P. G., E. J. Richardson, G. J. McKenzie, and A. N. McKenzie. 2000. Schistosome infection of transgenic mice defines distinct and contrasting pathogenic roles for IL-4 and IL-13: IL-13 is a profibrotic agent. *J. Immunol. 164:2585.*

29. La Flamme, A. C., E. A. Patton, B. Bauman, and E. J. Pearce. 2001. IL-4 plays a crucial role in regulating oxidative damage in the liver during schistosomiasis. *J Immunol 166:1903.*

30. La Flamme, A. C., E. A. Patton, and E. J. Pearce. 2001. Role of gamma interferon in the pathogenesis of severe schistosomiasis in interleukin-4-deficient mice. *Infect Immun 69:7445.*

31. Patton, E. A., L. R. Brunet, A. C. La Flamme, J. Pedras-Vasconcelos, M. Kopf, and E. J. Pearce. 2001. Severe schistosomiasis in the absence of interleukin-4 (IL-4) is IL-12 independent. *Infect Immun 69:589.*

32. Metwali, A., D. Elliott, A. M. Blum, J. Li, M. Sandor, R. Lynch, N. Noben-Trauth, and J. V. Weinstock. 1996. The granulomatous response in murine schistosomiasis mansoni does not switch to Th1 in IL-4-deficient C57BL/6 mice. *J Immunol 157:4546.*

33. Pearce, E. J., A. Cheever, S. Leonard, M. Covalesky, R. Fernandez-Botran, G. Kohler, and M. Kopf. 1996. Schistosoma mansoni in IL-4-deficient mice. *Int Immunol 8:435.*

34. Wynn, T. A., A. W. Cheever, M. E. Williams, S. Hieny, P. Caspar, R. Kühn, W. Müller, and A. Sher. 1998. IL-10 regulates liver pathology in acute murine schistosomiasis mansoni but is not required for immune down-modulation of chronic disease. *J Immunol 160:5000.*

35. Hoffmann, K. F., T. C. McCarty, D. H. Segal, M. Chiaramonte, M. Hesse, E. M. Davis, A. W. Cheever, P. S. Meltzer, H. C. Morse, 3rd, and T. A. Wynn. 2001. Disease fingerprinting with cDNA microarrays reveals distinct gene expression profiles in lethal type 1 and type 2 cytokine-mediated inflammatory reactions. *Faseb J 15:2545.*

36. Brunet, L. R., M. Beall, D. W. Dunne, and E. J. Pearce. 1999. Nitric oxide and the Th2 response combine to prevent severe hepatic damage during Schistosoma mansoni infection. *J Immunol 163:4976.*

37. Gryseels, B., and S. J. de Vlas. 1996. Worm burdens in schistosome infections. *Parasitology Today 12:115.*

38. Mwinzi, P. N., D. M. Karanja, D. G. Colley, A. S. Orago, and W. E. Secor. 2001. Cellular immune responses of schistosomiasis patients are altered by human immunodeficiency virus type 1 coinfection. *J Infect Dis 184:488.*

39. Cheever, A. W., and Z. A. Andrade. 1967. Pathological lesions associated with Schistosoma mansoni infection in man. *Trans R Soc Trop Med Hyg 61:626.*

40. Booth, M., J. K. Mwatha, S. Joseph, F. M. Jones, H. Kadzo, E. Ireri, F. Kazibwe, J. Kemijumbi, C. Kariuki, G. Kimani, J. H. Ouma, N. B. Kabatereine, B. J. Vennervald, and D. W. Dunne. 2004. Periportal fibrosis in human Schistosoma mansoni infection is associated with low IL-10, low IFN-gamma, high TNF-alpha, or low RANTES, depending on age and gender. *J Immunol 172:1295.*

41. Henri, S., C. Chevillard, A. Mergani, P. Paris, J. Gaudart, C. Camilla, H. Dessein, F. Montero, N. E. Elwali, O. K. Saeed, M. Magzoub, and A. J. Dessein. 2002. Cytokine Regulation of Periportal Fibrosis in Humans Infected with Schistosoma mansoni: IFN-gamma Is Associated with Protection Against Fibrosis and TNF-alpha with Aggravation of Disease. *J Immunol 169:929.*

42. Mwatha, J. K., G. Kimani, T. Kamau, G. G. Mbugua, J. H. Ouma, J. Mumo, A. J. Fulford, F. M. Jones, A. E. Butterworth, M. B. Roberts, and D. W. Dunne. 1998. High levels of TNF, soluble TNF receptors, soluble ICAM-1, and IFN- gamma, but low levels of IL-5, are associated with hepatosplenic disease in human schistosomiasis mansoni. *J Immunol 160:1992.*

43. Chiaramonte, M. G., D. D. Donaldson, A. W. Cheever, and T. A. Wynn. 1999. An IL-13 inhibitor blocks the development of hepatic fibrosis during a T-helper type 2-dominated inflammatory response. *J Clin Invest 104:777.*

44. Oriente, A., N. S. Fedarko, S. E. Pacocha, S. K. Huang, L. M. Lichtenstein, and D. M. Essayan. 2000. Interleukin-13 modulates collagen homeostasis in human skin and keloid fibroblasts. *J Pharmacol Exp Ther 292:988.*

45. Wynn, T. A. 2003. IL-13 effector functions. *Annu Rev Immunol 21:425.*

46. Vaillant, B., M. G. Chiaramonte, A. W. Cheever, P. D. Soloway, and T. A. Wynn. 2001. Regulation of hepatic fibrosis and extracellular matrix genes by the Th response: New insight into the role of tissue inhibitors of matrix metalloproteinases. *J. Immunol 157:00.*

47. Chiaramonte, M. G., A. W. Cheever, J. D. Malley, D. D. Donaldson, and T. A. Wynn. 2001. Studies of murine schistosomiasis reveal interleukin-13 blockade as a treatment for established and progressive liver fibrosis. *Hepatology 34:273.*

48. Stadecker, M. J., J. K. Kamisato, and S. M. Chikunguwo. 1990. Induction of T helper cell unresponsiveness to antigen by macrophages from schistosomal egg granulomas. A basis for immunomodulation in schistosomiasis? *J Immunol 145:2697.*

49. Flores Villanueva, P. O., T. S. Harris, D. E. Ricklan, and M. J. Stadecker. 1994. Macrophages from schistosomal egg granulomas induce unresponsiveness in specific cloned Th-1 lymphocytes in vitro and down-regulate schistosomal granulomatous disease in vivo. *J Immunol 152:1847.*

50. Flores Villanueva, P. O., H. Reiser, and M. J. Stadecker. 1994. Regulation of T helper cell responses in experimental murine schistosomiasis by IL-10. Effect on expression of B7 and B7-2 costimulatory molecules by macrophages. *J Immunol 153:5190.*

51. Sher, A., D. Fiorentino, P. Caspar, E. Pearce, and T. Mosmann. 1991. Production of IL-10 by CD4$^+$ T lymphocytes correlates with down-regulation of Th1 cytokine synthesis in helminth infection. *J.Immunol. 147:2713.*

52. Stadecker, M. J. 1999. The regulatory role of the antigen-presenting cell in the development of hepatic immunopathology during infection with Schistosoma mansoni. *Pathobiology 67:269.*

53. Bosshardt, S. C., G. L. Freeman, Jr., W. E. Secor, and D. G. Colley. 1997. IL-10 deficit correlates with chronic, hypersplenomegaly syndrome in male CBA/J mice infected with Schistosoma mansoni. *Parasite Immunol 19:347.*

54. Estaquier, J., M. Marguerite, F. Sahuc, N. Bessis, C. Auriault, and J. C. Ameisen. 1997. Interleukin-10-mediated T cell apoptosis during the T helper type 2 cytokine response in murine Schistosoma mansoni parasite infection. *Eur Cytokine Netw 8:153.*

55. Sadler, C. H., L. I. Rutitzky, M. J. Stadecker, and R. A. Wilson. 2003. IL-10 is crucial for the transition from acute to chronic disease state during infection of mice with Schistosoma mansoni. *Eur J Immunol 33:880.*

56. Henderson, G. S., N. A. Nix, M. A. Montesano, D. Gold, G. L. Freeman, Jr., T. L. McCurley, and D. G. Colley. 1993. Two distinct pathological syndromes in male CBA/J inbred mice with chronic Schistosoma mansoni infections. *Am J Pathol 142:703.*

57. Montesano, M. A., G. L. Freeman, Jr., W. E. Secor, and D. G. Colley. 1997. Immunoregulatory idiotypes stimulate T helper 1 cytokine responses in experimental Schistosoma mansoni infections. *J Immunol 158:3800.*

58. Montesano, M. A., D. G. Colley, S. Eloi-Santos, G. L. Freeman, Jr., and W. E. Secor. 1999. Neonatal idiotypic exposure alters subsequent cytokine, pathology, and survival patterns in experimental Schistosoma mansoni infections. *J Exp Med 189:637.*

59. Montesano, M. A., D. G. Colley, M. T. Willard, G. L. Freeman, Jr., and W. E. Secor. 2002. Idiotypes Expressed Early in Experimental Schistosoma mansoni Infections Predict Clinical Outcomes of Chronic Disease. *J Exp Med 195:1223.*

60. Cheever, A. W., J. E. Byram, S. Hieny, F. von Lichtenberg, M. N. Lunde, and A. Sher. 1985. Immunopathology of Schistosoma japonicum and S. mansoni infection in B cell depleted mice. *Parasite Immunol 7:399.*

61. Jankovic, D., A. W. Cheever, M. C. Kullberg, T. A. Wynn, G. Yap, P. Caspar, F. A. Lewis, R. Clynes, J. V. Ravetch, and A. Sher. 1998. CD4+ T cell-mediated granulomatous

pathology in schistosomiasis is downregulated by a B cell-dependent mechanism requiring Fc receptor signaling. *J Exp Med 187:619.*

62. Hernandez, H. J., Y. Wang, and M. J. Stadecker. 1997. In infection with Schistosoma mansoni, B cells are required for T helper type 2 cell responses but not for granuloma formation. *J Immunol 158:4832.*

63. Lundy, S. K., S. P. Lerman, and D. L. Boros. 2001. Soluble egg antigen-stimulated T helper lymphocyte apoptosis and evidence for cell death mediated by FasL(+) T and B cells during murine Schistosoma mansoni infection. *Infect Immun 69:271.*

64. Lundy, S. K., and D. L. Boros. 2002. Fas ligand-expressing B-1a lymphocytes mediate CD4(+)-T-cell apoptosis during schistosomal infection: induction by interleukin 4 (IL-4) and IL-10. *Infect Immun 70:812.*

65. Rutitzky, L. I., G. A. Mirkin, and M. J. Stadecker. 2003. Apoptosis by neglect of CD4+ Th cells in granulomas: a novel effector mechanism involved in the control of egg-induced immunopathology in murine schistosomiasis. *J Immunol 171:1859.*

66. Wood, N., M. J. Whitters, B. A. Jacobson, J. Witek, J. P. Sypek, M. Kasaian, M. J. Eppihimer, M. Unger, T. Tanaka, S. J. Goldman, M. Collins, D. D. Donaldson, and M. J. Grusby. 2003. Enhanced interleukin (IL)-13 responses in mice lacking IL-13 receptor alpha 2. *J Exp Med 197:703.*

67. Chiaramonte, M. G., M. Mentink-Kane, B. A. Jacobson, A. W. Cheever, M. J. Whitters, M. E. Goad, A. Wong, M. Collins, D. D. Donaldson, M. J. Grusby, and T. A. Wynn. 2003. Regulation and function of the interleukin 13 receptor alpha 2 during a T helper cell type 2-dominant immune response. *J Exp Med 197:687.*

68. Mentink-Kane, M. M., A. W. Cheever, R. W. Thompson, D. M. Hari, N. B. Kabatereine, B. J. Vennervald, J. H. Ouma, J. K. Mwatha, F. M. Jones, D. D. Donaldson, M. J. Grusby, D. W. Dunne, and T. A. Wynn. 2004. IL-13 receptor alpha 2 down-modulates granulomatous inflammation and prolongs host survival in schistosomiasis. *Proc Natl Acad Sci U S A 101:586.*

69. Hesse, M., A. W. Cheever, D. Jankovic, and T. A. Wynn. 2000. NOS-2 mediates the protective anti-inflammatory and antifibrotic effects of the Th1-inducing adjuvant, IL-12, in a Th2 model of granulomatous disease. *Am J Pathol 157:945.*

70. Hesse, M., M. Modolell, A. C. La Flamme, M. Schito, J. M. Fuentes, A. W. Cheever, E. J. Pearce, and T. A. Wynn. 2001. Differential Regulation of Nitric Oxide Synthase-2 and Arginase-1 by Type 1/Type 2 Cytokines In Vivo: Granulomatous Pathology Is Shaped by the Pattern of L-Arginine Metabolism. *J Immunol 167:6533.*

71. Modolell, M., I. M. Corraliza, F. Link, G. Soler, and K. Eichmann. 1995. Reciprocal regulation of the nitric oxide synthase/arginase balance in mouse bone marrow-derived macrophages by TH1 and TH2 cytokines. *Eur J Immunol 25:1101.*

72. Gordon, S. 2003. Alternative activation of macrophages. *Nat Rev Immunol 3:23.*

73. Dessein, A. J., D. Hillaire, N. E. Elwali, S. Marquet, Q. Mohamed-Ali, A. Mirghani, S. Henri, A. A. Abdelhameed, O. K. Saeed, M. M. Magzoub, and L. Abel. 1999. Severe hepatic fibrosis in Schistosoma mansoni infection is controlled by a major locus that is closely linked to the interferon-gamma receptor gene. *Am J Hum Genet 65:709.*

74. van der Pouw Kraan, T. C., A. van Veen, L. C. Boeije, S. A. van Tuyl, E. R. de Groot, S. O. Stapel, A. Bakker, C. L. Verweij, L. A. Aarden, and J. S. van der Zee. 1999. An IL-13 promoter polymorphism associated with increased risk of allergic asthma. *Genes Immun 1:61.*

75. Ahmed, S., K. Ihara, Y. Sasaki, F. Nakao, S. Nishima, T. Fujino, and T. Hara. 2000. Novel polymorphism in the coding region of the IL-13 receptor alpha' gene: association study with atopic asthma in the Japanese population. *Exp Clin Immunogenet 17:18.*

Chapter 10

IMMUNOLOGICAL AND OTHER FACTORS AFFECTING HEPATOSPLENIC SCHISTOSOMIASIS MANSONI IN MAN

David W. Dunne and Mark Booth
Division of Microbiology & Parasitology, Department of Pathology, University of Cambridge, Tennis Court Road, Cambridge, CB2 1QP, U.K.

Key words: hepatosplenic schistosomiasis, periportal fibrosis, IFN , TNF , malaria

1.   HEPATOSPLENIC SCHISTOSOMIASIS AND PERIPORTAL FIBROSIS

Hepatosplenic schistosomiasis (HS) is a severe clinical syndrome that can result from chronic *Schistosoma mansoni* or *S. japonicum* infections. HS presents as firm/hard splenomegaly or hepatosplenomegaly, often accompanied by portal hypertension with associated development of collateral circulation, gastro-eosphageal varices and sometimes, fatal haematemesis. This clinical picture is not unique to schistosomiasis, as it is a common outcome in hepatic lesions that cause significant constriction of the pre-sinusoidal portal blood flow, including cirrhosis, chemical hepatotoxicity, and bilinary carcinoma. What is unique to chronic infection with these schistosome species is Symmers' clay-pipestem fibrosis, a macroscopic fibrosis of the hepatic portal tract that was first reported from autopsies in Egypt a hundred years ago (Symmers, 1904). The clinical syndrome and the gross hepatic morbidity were causally associated by an important series of autopsy studies in Brazil (Cheever & Andrade 1967) and supportive observations in Egypt (Kamel et al. 1978). Cheever and colleagues found that, in *S. mansoni*-infected individuals, signs of portal hypertension were only found in the presence of Symmers' periportal fibrosis, and were not associated with lesser fibrosis of small hepatic veins, or with other hepatic lesions such as cirrhosis.

In his original observations of periportal fibrosis, Symmers related the lesion to the presence of parasite eggs in hepatic tissues. However, it was difficult to identify how the underlying morbidity might be triggered by the presence of the parasite or, at the epidemiological level, if the disease might be influenced by interactions with environmental and other risk factors. Autopsy studies, by their nature, focus on a highly non-randomly selected cohort. Moreover, methods able to identify periportal fibrosis in the context of epidemiological or community studies did not exist and clinical procedures, such as hepatic wedge and needle biopsies, were both highly invasive and unsuited for the detection and characterization of gross periportal fibrosis. It is only since the general availability of portable ultrasound machines, in the last twenty years, that it has been possible to relate clinical HS and portal tract fibrosis in infected communities. Characteristic schistosomiasis-associated mild to severe periportal fibrosis can now easily be visualized and categorized using portable ultrasound. In addition, ultrasonographic measurements of portal and splenic vein diameters can indicate increased portal pressure, and gastro-oesophgeal varices can be detected with a useful degree of sensitivity. Liver and spleen size measurements can also be made using ultrasound. However, additional physical clinical examination still has some advantages in that the degree of palpable organ enlargement below the costal arc does not need to be related to a height-standardized normogram and can provide important information as to whether the palpable organs have a soft or firm/hard consistency.

2.   EPIDEMIOLOGICAL FACTORS AFFECTING THE PREVALENCE OF PERIPORTAL FIBROSIS

The application of ultrasonography to the detection of schistosomiasis-associated hepatic periportal fibrosis within populations living in *S. mansoni* endemic areas has revealed that the prevalence and severity of this fibrosis peaks relatively late in life (Mohamed-Ali et al. 1999; Kariuki et al. 2001). Although it is often difficult to separate the effects of age, duration and cumulative amount of exposure in schistosomiasis endemic areas, a recent series of epidemiological studies has managed to do this successfully. The studies took place in neighboring fishing communities on the Ugandan shore of Lake Albert in the Albertine Rift valley that form the western border of that country. Collaborative work in one parish, Butiaba, has been ongoing since 1996, and several communities have now been surveyed and treated.

In common with many other studies, the average number of *S. mansoni* eggs in stool samples in each community followed a convex profile with age, with peak intensities of infection observed in children around the age of 12 to 14 years. In the past, interpretation of these patterns has been complicated by observations that children are often more exposed than adults.   In Uganda, however, the confounding question was resolved by observations that adults in the fishing village of Piida had considerably more contact with water than children (Kabatereine et al. 1999) due to occupational activities.

The lower infection intensities combined with increased exposure amongst adults provided evidence of resistance to infection amongst older inhabitants despite an increasing risk of periportal fibrosis. To understand more about this three-way relationship, an epidemiological study was undertaken of two cohorts from neighboring lakeside fishing villages, called Booma and Bugoigo. Cohort members were selected on the basis of having been resident in the area for at least ten years. This was an important criterion since it prevented recent migrants with an uncertain history of exposure possibly confounding the analysis. Each member of the cohort was asked about their duration of residence, and examined by ultrasound for evidence of periportal fibrosis (Figure 10-1a) using the semi-quantitative Niamey image patterns (Figure 10-1b) as defined by WHO criteria (Richter et al. 2000; Chippaux et al. 2000).   A striking observation was that the prevalence of periportal fibrosis in Booma was 31.5% whereas in the Bugoigo cohort the prevalence was significantly lower at 6.1% - a 5 fold increased probability of fibrosis amongst long-term residents of Booma.

The difference was unlikely to be due to any differences in genetic background or age since the demographic make-up of the two cohorts was similar, with the two main tribes in the area (Mugungu and Alur) being strongly represented in each community. Furthermore, the prevalence of infection was 92.5% and 87.9% in the Booma and Bugoigo cohorts respectively, and mean egg counts were only slightly lower in Bugoigo. These parasitological observations indicated that transmission levels were similar in the two villages. A major difference was that, despite the selection criterion of long-term residency, the inhabitants of Booma had been resident for considerably longer than those of Bugoigo. This was true of each age group. When the data from the two cohorts were pooled, it became clear that the risk and severity of fibrosis increased with duration of residency (Figure 10-1c). The most dramatic difference was observed within the 17-32 year age group, where the probability of fibrosis was 12 fold higher amongst those resident 22-31 years compared with those who had been resident for between 10 and 15 years (Booth et al, 2004a). Interestingly, the differences in risk of fibrosis were associated with decades of difference in duration of residency, suggesting that periportal fibrosis amongst the Bantu (Mugunga) and Nilotic (Alur) peoples is the result of cumulative insults to the liver over many years, even in the absence of massive egg excretion.

a)   **b)**   **c)**

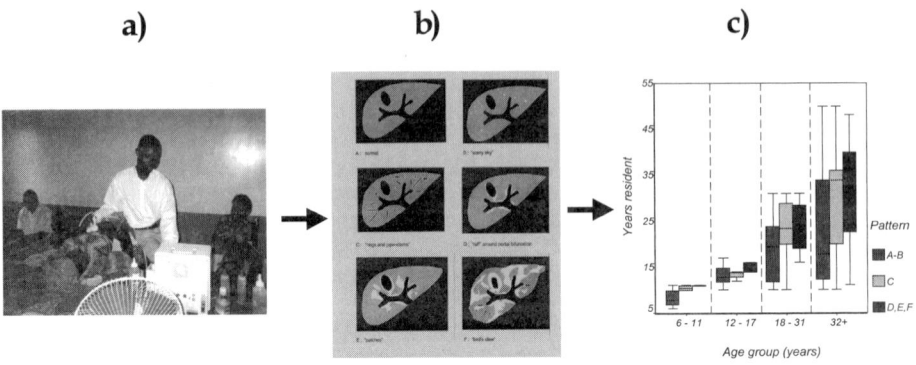

Figure 10-1. In Uganda, the level of periportal fibrosis was recorded semi-quantitatively using a portable ultrasound machine (Figure 10-1a). The scoring was based on the nature and extent of fibrotic tissue in the liver (Figure 10-1b). These pattern scores were then used in comparing years of residency in the area versus age on the risk of fibrosis (Figure 10-1c).

3.   IMMUNOLOGICAL FACTORS ASSOCIATED WITH FIBROSIS

A major advance towards identifying the underlying cause of hepatosplenic schistosomiasis stemmed from Ken Warren's hypothesis, based on the mouse model of schistosomiasis mansoni, that hepatosplenic schistosomiasis had an immunological basis (Warren 1975). Since then the mouse model has proved to be an invaluable tool for studying the immunological aspects of hepatosplenic disease. Although this experimental host does not develop periportal fibrosis, it has been instrumental in the great progress that has been made in understanding the regulation and counter-regulation of the host granulomatous response to the schistosome egg and intra-granuloma fibrosis. It is now very well established in the mouse model that egg-induced Th2 cytokine responses such as IL-4 and, particularly, IL-13, promote murine hepatic fibrosis, and that this is counter-regulated by IL-12 and IFNγ (Hoffmann et al. 2002).

The extremely late on-set of schistosomiasis-associated periportal fibrosis in man, often at a stage when peak intensities of infection have passed, has made it very difficult to relate human anti-parasite immune responses with the chronic disease. Nonetheless, the high prevalence of periportal fibrosis observed in Booma provided an opportunity to examine the immunological factors that may contribute to the etiology of the disease. An interesting observation was that the overall prevalence of fibrosis in adult women was

25.8%, which was significantly lower than the prevalence of 60.3% observed in men. Women had similar infection levels, and had not been resident for any shorter time than the men. Water contact data from the neighboring Piida community suggested that one reason for the lower prevalence of fibrosis in women was due to a lower number of hours contact with infested lake water (Kabatereine et al. 1999). However, subsequent analysis of immunological assay data (IL-1β, IL-3, IL-4, IL-5, IL-10, IL-13, TNFα, IFNγ, RANTES each measured constitutively and in response to 3 crude antigens) gave rise to another possibility – that sex-specific immunological responses may also be important (Booth et al, 2004[b]).

To make the dataset more manageable, factor analysis was first used to reduce 36 variables to seven factors, with each factor representing several correlated assays. A risk analysis based on recursive partitioning algorithms (Zhang & Singer 1999) was then used to identify combinations of demographic and immunological factors associated with either a high or low risk of fibrosis. It was possible to identify broad trends as well as to identify smaller clusters of cases with particularly high or low risks that corresponded to a specific combination of factors.

Gender and age were identified as the most important risk factors; therefore children, adult males and adult females were analyzed separately. Within these groups it emerged that 8 from 9 children with fibrosis had low IL-10 levels. Amongst women, the highest risk of fibrosis (44%) was associated with a relatively high TNFα response. We also observed a group of ten women with the highest levels of IFNγ amongst whom there were no cases. Amongst men, a high risk of fibrosis (>90%) was associated with either a low level of RANTES or elevated TNFα. Importantly, these observations concurred with an earlier study of severe fibrosis in the Sudan, where there was no sex stratification, but elevated IFNγ was associated with a low risk and elevated TNF with a high risk of ultrasound-detectable fibrosis (Henri et al. 2002). A genetic link to elevated production of IFNγ had already been identified (Dessein et al. 1999). Taken together, these results suggest that the risk of periportal fibrosis in humans is correlated with sustained exposure to the infection, but is also associated at the individual level with an innate ability to produce certain cytokines, particularly IFNγ and TNFα.

The association of higher IFNγ with reduced risk of periportal fibrosis suggests that it has an anti-fibrosis role in human schistosomiasis (Dessein et al. 1999). It is known that IFNγ is able to inhibit fibroblast proliferation and matrix production making it a powerful antagonist of fibrogenesis (Duncan & Berman 1985; Ghosh et al. 2001). In schistosome-infected mice hepatic fibrosis has been associated with increased expression of TNFα and it has been shown that treatment with IFNγ inhibited collagen deposition (Czaja et al. 1989). During inflammation, increased levels of TNFα and soluble forms of its receptors are up-regulated during inflammatory responses (Kuhns et al. 1995). TNFα activates macrophages to produce oxidants that cause toxicity and destruction of endothelial lining of blood vessels (Slungaard et al. 1990). Human *in vitro* production and circulating levels of IFNγ and TNFα (or sTNFR-II), respectively, have now been associated with the absence and presence of schistosomiasis mansoni ultrasound-detectable periportal fibrosis in the Sudan (Henri et al. 2002) and Uganda (Booth et al. 2004[b]). High plasma levels of biologically active TNF have also been associated with schistosomiasis mansoni hepatosplenic disease in Brazil, although periportal fibrosis was not formally reported in this study (Zwingenberger et al. 1990). Thus, the association of high levels of TNF and IFNγ with, respectively, greater and lesser risk of periportal fibrosis, appears to be approaching a consensus in relation to mouse studies and human populations in different *S. mansoni* endemic areas.

There is little information on the roles of RANTES in human schistosomiasis morbidity. IFNγ and to a lesser extent TNFα, can induce RANTES production (Berkman et al. 1996; Katayama et al. 2002), and it has been suggested that, not only are RANTES and IFNγ co-expressed, but they can also function synergistically (Dorner et al. 2002). RANTES has been shown to be involved in Th2 allergy models, such as the accumulation of eosinophils in the lungs (Lukacs 2000). However, in a study of Th1 *Mycobacteria bovis* granuloma and Th2 *S. mansoni* granuloma in the mouse, RANTES dominated in the Th1 response, but inhibited the Th2 response and, unlike in allergy models, RANTES prevented the accumulation of eosinophils in *S. mansoni* granulomas (Chensue et al. 1999). The association of low levels of RANTES with periportal fibrosis in *S. mansoni*-infected Ugandans now suggests a further regulatory role for this chemokine.

4.   HS IN THE ABSENCE OF PERIPORTAL FIBROSIS

For many years, before the advent of portable ultrasound equipment, classical epidemiology studies relied on the presence and extent of hepatosplenomegaly to classify the severity of *S. mansoni* hepatic morbidity. Such studies established that hepatosplenomegaly coincided with peak intensities of *S. mansoni* infections, as measured by parasite egg excretion, in late childhood (Lehman et al. 1976; Gryseels & Polderman 1987). Although some studies have reported greater intensities of infection in adults suffering periportal fibrosis compared with those who do not, many others, like Booth and colleagues (2004a) have observed the greatest prevalence and severity of periportal fibrosis is often found in older individuals well after the peak age of infection intensity when little or no schistosome infection is detectable (Homeida et al. 1988). In addition, combined clinical and ultrasonography studies in *S. mansoni* endemic areas have reported that hepatosplenomegaly can occur in the absence of detectable periportal fibrosis. In a study in three villages of Madagascar, where *S. mansoni* is hyperendemic, Boisier and colleagues (2001) reported ultrasonographic findings typical for portal hypertension in the absence of any detectable periportal fibrosis in 4.3% of individuals. In an epidemic focus of *S. mansoni* infection in Senegal, three years after the recorded start of the outbreak, only a few infected-subjects had developed slight left lobe hepatomegaly at clinical examination. Three years later, the overall prevalence of hepatosplenomegy had increased to 36%, with 49% of 5 to 19 year old *S. mansoni*-infected subjects affected. Hepatosplenomegaly increased from 0 to 7%, and was associated with the heaviest *S. mansoni* infections (Stelma et al. 1997). Ultrasound examinations found no significant periportal fibrosis in these communities (Kardoff et al. 1996). These observations in Senegal also show that hepatosplenomegaly, even in the absence of detectable periportal fibrosis, takes a number of years to develop in communities exposed to high transmission of *S. mansoni* and, thus, is distinct from previously described (Rabello 1995) 'acute' forms of schistosomiasis.

Although schistosomiasis mansoni-associated clinical HS has been widely reported, the prevalence of this morbidity varies between different regions. The intensity of *S. mansoni* transmission is clearly an important factor associated with high prevalence of clinical HS (Ongom & Bradley, 1972; Arap Siongok et al. 1976; Smith et al. 1979; Roux et al. 1980; Gryseels & Polderman 1987), however we have identified areas in Kenya that have contrasting prevalence of HS despite similar infection intensities (Butterworth et al. 1989; Butterworth et al. 1994). In Kangundo no more than 0.5% of infected schoolchildren were found to have schistosomiasis-associated HS, whereas Kambu had many patients with severe HS. A comparative study of *S. mansoni* infected school children (6 and 19 years) from both areas shows that there was a positive relationship between infection intensity and palpable hepatosplenomegaly. Although the overall prevalence and mean intensities of infection were comparable, prevalence of schistosome-associated HS was much higher in Kambu (Fulford et al. 1991; Butterworth et al. 1991; Corbett et al. 1992). The hepatosplenomegaly in Kambu peaked between 11 and 16 years of age, afflicting ~20% of school children, was frequently massive, and in the most severely affected children endoscopy revealed esophageal varices.

5.   IMMUNOLOGICAL CORRELATES OF CHILDHOOD HS

In a case-control study of children suffering clinically defined hepatosplenomegaly from Kambu and Kangando, each hepatosplenic child in Kambu was matched with two non-HS controls, one from each area, of similar ages and infection intensities. After *in vitro* stimulation with soluble worm antigen (SWA) or SEA, PBMC from HS cases responded to with significantly higher IFNγ and TNF, but lower IL-5, compared to matched non-HS controls, and Principle Component analysis identified a high IFNγ/TNF and low IL-5 axis in the data as the first principle component, which was significantly associated with HS. High levels of sTNFR-I, sTNFR-II and soluble Intercellular Adhesion Molecule 1 (sICAM-1) were also significantly associated with HS. Levels of all these were negatively correlated to IL-5 and positively to IFNγ (Mwatha et al. 1998). These results, combined with the known regulatory interactions between these molecules, suggested that in HS children, low Th2 cytokine responses and corresponding high IFNγ, leads to up-regulation of TNF in IFNγ activated macrophages and other cells. Increased hepatic TNF would up-regulate expression of adhesion molecules involved in the aggregation of inflammatory cells. Therefore, increased plasma sTNFR and sICAM-1 could be markers of chronic proinflammatory activity in hepatic tissues. As recent studies have confirmed that this form of childhood HS in Kenya is not dependent on the presence of periportal fibrosis (Vennervald et al. In press [b]) the presence of hepatosplenomegaly and increased portal pressure, suggests that TNF- and IFNγ-dependent inflammatory processes may be constricting hepatic portal blood flow. Stelma and colleagues (1997), when confronted with similar clinical observations in Senegal, speculated that large numbers of hepatic egg granulomas might be responsible. This could be the case, as TNFα is a crucial mediator for *S. mansoni* egg granuloma formation in the mouse liver (Amiri et al. 1992) and the failure to down-regulate TNFα has been associated with a hepatosplenomegaly syndrome in *S. mansoni* infected mice (Adewusi et al. 1996).

6.   CO-INFECTIONS AND HS - SCHISTOSOMIASIS AND MALARIA INTERACTIONS?

Clinical HS defined largely by firm/hard hepatosplenomegaly is, like periportal fibrosis, unevenly distributed across schistosomiasis mansoni endemic areas. Although high intensity of infection is a clear risk factor, the strikingly different prevalence of HS in areas of similar parasite transmission points to the existence of additional important risk factors. The identification of the differences in the prevalence of hepatosplenomegaly in Akamba schoolchildren from Kambu and Kangundo allowed Butterworth and colleagues to investigate a number of epidemiological and other factors. *S.mansoni* strain differences; differences in the patterns of exposure to infection; host nutritional status; exposure to other infectious or toxic agents, including *S. haematobium*, hookworm, *Ascaris, Trichuris, Enterobius, Entamoeba histolytica, Entamoeba coli, Giardia lamblia, Brucella abortus* and malaria, were all investigated (Fulford et al. 1991; Corbett et al. 1992; Chunge et al. 1995). None provided a convincing explanation. Children in Kambu were nutritionally disadvantaged compared with those of Kangundo, and of the infectious agents, only malaria was found to be more prevalent. However, even though HS in Kambu was not associated with the presence of blood smear-detectable malaria infections, the possibility that some aspect of greater exposure to malaria might be contributing to the high prevalence for hepatosplenomegaly in *S. mansoni*-infected children could not be eliminated (Fulford et al. 1991).

Malaria can cause hepatosplenomegaly in the absence of schistosome infection. However, malaria organomegaly is more often found in much younger pre-school age children and the enlarged organs, which usually have a soft consistency, regress rapidly with successful treatment. Nonetheless, in Africa, *S. mansoni* and malaria are often co-endemic, and morbidity studies of either infection would do well to consider the potentially confounding effects of the other infection (Omer et al. 1976). Buck and colleagues (1978) carried out studies on polyparasitism in villages in Chad that were hyperendemic for schistosomiasis mansoni and malaria or malaria only and concluded that the combination of hyperendemic malaria and heavy *S. mansoni* infection resulted in a higher prevalence of hepatomegaly and splenomegaly than either malaria or schistosomiasis alone. In a study conducted in former Rhodesia, associations between hepatosplenomegaly, schistosomiasis, malaria and malnutrition were examined in children from an area endemic for schistosomiasis and malaria. In the 0 to 4 year olds, both liver and spleen enlargement were significantly associated with malaria, whereas in the 5 to 9 year old children hepatomegaly was found to be significantly associated with *S. mansoni* infection, and with combined *S. mansoni* and malaria infection (Whittle *et al.* 1969). In villages in Zaire, having similar prevalence of malaria but different *S. mansoni* transmission (Gryseels & Polderman, 1987), a 26% overall prevalence of hepatosplenomegaly (highest 6 to 18 year-olds) was found where *S. mansoni* transmission was high, with much lower organomegaly where transmission was low. In the same study, hepatomegaly positively correlated with high infection intensities of schistosomiasis in children but not in adults. The authors speculated that perhaps organomegaly in the adults, but not children, was associated with Symmers' fibrosis. Stelma and colleagues (1997) also wondered if malaria might be contributing to the high prevalence of hepatosplenomegaly that they observed in *S. mansoni*-infected subjects in the absence of detectable periportal fibrosis.

We retrospectively assayed plasma from the Kambu/Kangundo case-control study (Mwatha et al. 2003) for anti-*P. falciparum* schizont antibodies to assess whether or not they provide evidence that exposure to *Plasmodium* may be a factor in the development of hepatosplenomegaly in *S. mansoni*-infected children. Anti-malarial antibody responses, particularly IgG3, were generally higher in Kambu than in Kangundo. However, more interesting was the observation that there was a difference in anti-malaria antibody responses between Kambu cases and Kambu controls, with hepatosplenic cases having markedly higher anti-malaria responses than area-matched controls. Thus, possession of anti-malarial antibodies appears to be a major risk factor for the development of hepatosplenomegaly in schistosome-infected schoolchildren living in this co-endemic area. This strongly indicated that some aspect of exposure to malaria influences the development of hepatosplenomegaly in schistosomiasis-infected children in parts of Kenya. Again this did not depend on either acute clinical malaria or patent malaria parasitaemia; however, chronic exposure or chronic sub-patent malaria infection may be involved. We have recently found in children from the Kambu area who were blood smear negative for malaria, who nevertheless had PCR detectable malaria in their blood, had higher levels of IgG3 anti-schizont responses compared with those who were negative by both blood smear and PCR (unpublished results). This suggests that high anti-schizont IgG3 may reflect some combination of chronic low level malaria infection, or of chronic exposure to malaria.

7.   INVESTIGATING THE EFFECT OF CO-INFECTIONS ON HS USING GIS.

The above studies could only hint at the role of malaria in schistosome associated HS, since they were not designed to take into account the seasonal aspects of malaria transmission. Since all the surveys were conducted outside the transmission season, it was not surprising to find no correlation between malaria parasitaemia and HS (Fulford et al. 1991; Mwatha et al. 2003). However, had the surveys been conducted during the malaria transmission season, transient inflammation of both liver and spleen associated with acute malaria could have confounded attempts to estimate the role of *S. mansoni* infection. A multidisciplinary approach has proved valuable in tackling this problem during a longitudinal study of HS, conducted over three years, amongst a cohort of school-aged children living in the area of Kambu (now in Makueni district), Kenya (Vennervald et al. In press [a, b]). The 96 children were selected on the criterion of hepatomegaly detectable by ultrasonography, based on a Senegalese normogram (Yazdanpanah et al. 1997), and most of them also presented with clinically detectable HS at the time of the pre-treatment examination, which took place well outside the malaria transmission season. Whereas variation in the degree of hepatomegaly was very limited due to the selection process, there was considerable variation in the degree of splenomegaly. A weak but positive correlation between *S. mansoni* infection intensity and splenomegaly was observed, implicating *S. mansoni* as one etiological agent, but as expected there was no correlation with malaria parasitaemia. There was, however, a strong positive correlation between IgG3 response to schizont antigen and splenomegaly, indicating that exposure to malaria may also have been influential (Kenty et al. In preparation).

GIS analysis of the 10x4km study area (Fig 10-2a) revealed that the highest *S. mansoni* egg counts were clustered amongst children living towards the west of the study area, whereas IgG3 responses to *P. falciparum* schizont antigen were clustered along the northern border and decreased with distance from the river (Figure 10-2b). These observations meant only limited overlap in the geographical distribution of relatively high exposure to both infections, with clearly defined areas corresponding to high or low level exposure to one or both infections. When the magnitude of splenomegaly was compared across these areas, it was observed that children living in the area with high egg counts and high IgG3 values had considerably bigger spleens than those living elsewhere. The inference was that a combination of relatively high exposure to both infections exacerbated the degree of splenomegaly significantly beyond that associated with relatively high exposure to either infection alone (Booth et al. 2004[c]).

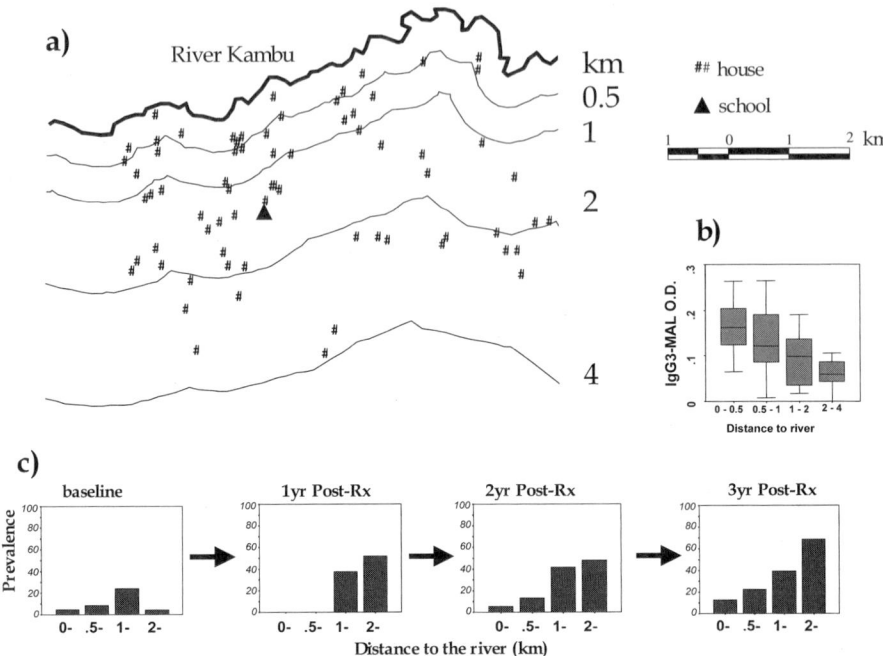

Figure 10-2. Panel a) shows the geographical distribution of houses within the Mbeetwani study area that is bounded by the River Kambu to the north. The lines across the map correspond to varying distances from the river. Figure 2b shows how the IgG3 responses against malaria antigen decreased with distance from the river. Figure 2c represents the trend for the prevalence of normal (non-palpable) spleens to increase both with time after praziquantel treatment, and distance from the river.

The longitudinal aspect of this study afforded an opportunity to assess the results of removing one infection (*S. mansoni*) on the regression of HS in the presence of continuing exposure to the other infection. Mollusciciding of the Kambu River over the next three years ensured limited re-infection of *S. mansoni*. Analysis of data collected at the same time-point each year revealed that splenomegaly decreased significantly (Vennervald et al. In press [b]). After three years, children from the area with relatively high transmission of both infections still had the largest spleens, although they were diminished when compared to baseline values. There was also a strong relationship between distance of domicile from the river and the probability of presenting with an un-palpable spleen (Figure 10-2c). Children from the area with relatively high exposure to malaria but low *S. mansoni* egg counts did not regress in terms of hepatomegaly. Together, these observations indicated that the level of exposure to malaria may not only exacerbate the severity of HS in school-aged children, but may also affect the outcome of praziquantel treatment on this widespread but neglected disease.

8.   CONCLUSIONS

Over 40 years ago, Walters & McGregor (1960) pointed out "Throughout the tropical belt hepatic disease is rife." Today the situation appears to have little changed. HS undoubtedly has multiple etiologies, and it is probably not uncommon that several will contribute to the clinical condition of individual patients. In our studies, we observe two distinct forms of HS associated with *S. mansoni* infections. Classical periportal fibrosis is accompanied by hepatosplenomegaly, or sometimes splenomegaly only, with increased portal pressure, or frank portal tension with gastro-esophageal varices. Severe cases may suffer from haematemesis and deaths do occur. Although this form of HS is sometimes seen in young children, its prevalence increases with age in endemic areas, and peaks usually after the age of peak infection intensities. However, length and magnitude of exposure to infection appears to be a more important risk factor than age per-se. We also observe HS in the absence of periportal fibrosis. The prevalence and severity of this increases with intensity of *S. mansoni* infection and is usually greatest in school-aged children. A characteristic sign is firm or hard hepatomegaly or hepatosplenomegaly, which can be massive. In severe cases, signs of portal hypertension and gastro-esophageal varices can be identified. Although more common in school-aged children, we have seen adult cases. The prevalence of childhood HS can vary greatly between different areas, even when the transmission of *S. mansoni* is similar. In the area around Kambu in Kenya, which has a very high prevalence of this form of HS, relatively high-level exposure to malaria is a major risk factor among school-aged children.

Periportal fibrosis is associated with high levels of *in vitro* TNF and low levels of IFNγ in response to parasite antigens. Low levels of RANTES are also associated with periportal fibrosis. HS in schoolchildren who have no periportal fibrosis detectable by ultrasonography, also have high TNF responses and high plasma sTNFR. However, in contrast to periportal fibrosis cases, this form of HS is associated with high levels of IFNγ. Thus, IFNγ appears to 'promote' childhood HS whilst 'protecting' against the development of later periportal fibrosis. These apparently opposing roles of IFNγ in the two forms of hepatic morbidity highlight the balance needed in the human host's attempt to manage the chronic effects of *S. mansoni* infection.

Although HS in the absence of periportal fibrosis takes a number of years to develop, it is an earlier event than the periportal fibrosis itself. Therefore, the relationship between HS and the risk of subsequent development of periportal fibrosis remains an important question. It has long been suspected that co-infection, and interactions with environment factors, may have an important affect on the prevalence and severity of many common morbidities in Sub-Saharan Africa. Recent observations, although tentative, suggest that the severity of childhood HS in particular may be influenced by both schistosome and malaria infections. It is also possible that the immune responses to schistosome antigens that are associated with childhood HS are skewed towards Th1 by the influence of malaria. A multidisciplinary approach, including clinical, ultrasound, immunological, demographic and, most recently, GIS methods, is proving highly informative in answering such questions in this important subject.

ACKNOWLEDGEMENTS

This work was carried out in collaboration with the Kenyan Medical Research Institute, the Division of Vector Borne Diseases of the Kenyan Ministry of Health, the Vector Control Division of the Ugandan Ministry of Health; and the Danish Bilharizis Laboratory. These collaborative studies were funded by the Wellcome Trust, the British Medical Research Council and The Commission of the European Community's, Science and Technology for Development Programme (INCO-DC contract IC18 CT97-0237 and INCO-DEV contract ICA4-CT-1999-10003).

REFERENCES

Adewusi, O.I., Nix, N.A., Lu, X., Colley, D.G., Secor, W.E. (1996) *Schistosoma mansoni*: relationship of tumor necrosis factor-alpha to morbidity and collagen deposition in chronic experimental infection. *Exp Parasitol* 84:115-123.

Amiri, P., Locksley, R.M., Parslow, T.G., Sadick, M., Rector, E., Ritter, D., McKerrow, J.H. (1992) Tumour necrosis factor alpha restores granulomas and induces parasite egg-laying in schistosome-infected SCID mice. *Nature* 356:604-607.

Arap Siongok, T.K. Mahmoud, A.A., Ouma, J.H., Warren, K.S., Muller, A.S., Handa, A.K., Houser, H.B. (1976) Morbidity in Schistosomiasis mansoni in relation to intensity of infection: study of a community in Machakos, Kenya. *Am J Trop Med Hyg* 25:274-284.

Berkman, N., Robichaud, A., Krishnan, V.L., Roesems, G., Robbins, R., Jose, P.J., Barnes, Chung, K.F. (1996) Expression of RANTES in human airway epithelial cells: effect of corticosteroids and interleukin-4, -10 and -13. *Immunology* 87:599-603.

Boisier, P., Ramarokoto, C-E., Ravoniarimbinina, P., Rabarijaona, L., Ravaoalimalala, V.E. (2001) Geographic differences in hepatosplenic complications of schistosomiasis mansoni and explanatory factors of morbidity. *Trop Med Int Health*, 6, 699-706.

Booth, M., Vennervald B.V., Kabatereine N.B., Kazibwe F., Ouma, J.H., Kariuki, C.H., Muchiri, E., Kadzo, H., Ireri, E., Kimani, G., Mwatha, J.K., Dunne, D.W. (2004a) Hepatosplenic morbidity in two neighboring communities in Uganda with high levels of *Schistosoma mansoni* infection but very different durations of residence. *Trans R Soc Hyg Trop Med.* 98:125-136.

Booth, M., Mwatha, J.K., Joseph, S., Jones, F.M., Kadzo, H., Ireri, E., Kazibwe, F., Kemijumbi, J., Kariuki, C.H., Kimani, G., Ouma, J.H., Kabatereine, N.B., Vennervald, B.J., Dunne, D.W. (2004b) Peri-portal fibrosis in human *Schistosoma mansoni* infection is associated with low IL-10, low IFNγ, high TNFα or low RANTES, depending on age and gender. *J Immunol* 172:1295-1303.

Booth, M., Vennervald, B.J., Kenty, L.C., Butterworth, A.E., Kariuki, C.H., Kadzo, H., Ireri, E., Amaganga, C., Gachuhi, K., Mwatha, J.K., Otedo, A., Ouma, J.H., Dunne, D.W. (2004c) Micro-geographical variation in exposure to *Schistosoma mansoni* and malaria affects splenomegaly amongst Kenyan school-aged children. BMC Infectious Diseases

Buck, A.A., Anderson, R.I., MacRae, A.A. (1978) Epidemiology of Poly-Parasitism. IV. Combined effects on the state of health. *Tropenmed Parasit* 29:253-68.

Butterworth, A.E., Corbett, E.L., Dunne, D.W., Fulford, A.J.C., Kimani, G., Gachuhi, K., Klumpp, R.K., Mbugua, G.G., Ouma, J.H., Siongok, T.K.A. and Sturrock, R.F. (1989) Immunity and morbidity in human schistosomiasis. In: *New Strategies in Parasitology*, Ed. McAdam, K.P.W.J. Churchill Livingstone, London, pp.193-210.

Butterworth, A.E., Sturrock, R.F., Ouma, J.H., Mbugua, G.G., Fulford, A.J.C., Kariuki, H.C. and Koech, D. (1991) Comparison of different chemotherapy strategies against *Schistosoma mansoni* in Machakos District, Kenya: effects on human infection and morbidity. *Parasitology* 103:339-355.

Butterworth, A.E., Curry, A.J., Dunne, D.W., Fulford, A.J.C., Gachuhi, K., Kariuki, H.C., Klumpp, R.K., Koech, D., Mbugua, G., Ouma, J.H., Roberts, M., Thiong'o, F.W., Capron, A. and Sturrock, R.F. (1994) Immunity and morbidity in human schistosomiasis mansoni. *Trop Geog Med* 46:197-208.

Cheever, A.W., Andrade, Z.A. (1967) Pathological lesions associated with *Schistosoma mansoni* infection in man. *Trans R Soc Trop Med Hyg* 61:626-639.

Chensue, S.W., Warmington, K.S., Allenspach, E.J., Lu, B., Gerard, C., Kunkel, S.L., Lukacs, N.W. (1999) Differential expression and cross-regulatory function of RANTES during mycobacterial (type 1) and schistosomal (type 2) antigen-elicited granulomatous inflammation. *J Immunol* 163:165-173.

Chippaux, J. P., Garba, A., Boulanger, D., Ernould, J.C., Engels, D. (2000) Reduced morbidity of schistosomiasis: report from an expert workshop on the control of schistosomiasis held at CERMES (15-18 February 2000, Niamey, Niger). *Bull Soc Pathol Exot* 93:356-360.

Chunge, R.N., Karumbra, P.N., Ouma, J.H., Thiong'o, F.W., Sturrock, R.F., Butterworth, A.E. (1995) Polyparasitism in two rural communities with endemic *Schistosoma mansoni* infection in Machakos District, Kenya. *J. Trop. Med. Hyg.* 98:440-444.

Corbett, E.L., Butterworth, A.E., Fulford, A.J.C., Ouma, J.H., Sturrock, R.F. (1992) Nutritional status of children with schistosomiasis mansoni in two different areas of Machakos District, Kenya. *Trans. Roy. Soc. Trop. Med. Hyg.* 86:266-273.

Czaja, M. J., Weiner, F.R., Takahashi, S., Giambrone, M.A., van der Meide, P.H., Schellekens, H., Biempica, L., Zern, M.A. (1989) Gamma-interferon treatment inhibits collagen deposition in murine schistosomiasis. *Hepatology* 10:795-800.

Dessein, A. J., Hillaire, D., Elwali, N.E., Marquet, S., Mohamed-Ali, Q., Mirghani, A., Henri, S., Abdelhameed, A.A., Saeed, O.K., Magzoub, M.M., Abel, L. (1999) Severe hepatic fibrosis in Schistosoma mansoni infection is controlled by a major locus that is closely linked to the interferon-gamma receptor gene. *Am J Hum Genet* 65:709-721.

Dorner, B. G., Scheffold, A., Rolph, M.S., Huser, M.B., Kaufmann, S.H., Radbruch, A., Flesch, I.E., Kroczek, R.A. (2002) MIP-1alpha, MIP-1beta, RANTES, and ATAC/lymphotactin function together with IFN-gamma as type 1 cytokines. *Proc Natl Acad Sci U S A* 99:6181-6186.

Duncan, M. R., Berman, B. (1985) Gamma interferon is the lymphokine and beta interferon the monokine responsible for inhibition of fibroblast collagen production and late but not early fibroblast proliferation. *J Exp Med* 162:516-527.

Fulford, A.J.C., Mbugua, G.G., Ouma, J.H., Kariuki, H.C., Sturrock, R.F. and Butterworth, A.E. (1991) Differences in the rate of hepatosplenomegaly due to *Schistosoma mansoni* infection between two areas in Machakos District, Kenya. *Trans. Roy. Soc. Trop. Med. Hyg.* 85:481-488.

Ghosh, A. K., Yuan, W., Mori, Y., Chen, S., Varga, J. (2001) Antagonistic regulation of type I collagen gene expression by interferon-gamma and transforming growth factor-beta. Integration at the level of p300/CBP transcriptional coactivators. *J Biol Chem* 276:11041-11048.

Gryseels, B., Polderman, A.M. (1987) The morbidity of schistosomiasis mansoni in Maniema (Zaire). *Trans R Soc Trop Med Hyg* 81:202-209.

Henri, S., Chevillard, C., Mergani, A., Paris, P., Gaudart, J., Camilla, C., Dessein, H., Montero, F., Elwali, N.E., Saeed, O.K., Magzoub, M., Dessein, A.J. (2002) Cytokine regulation of periportal fibrosis in humans infected with *Schistosoma mansoni*: IFN-gamma is associated with protection against fibrosis and TNF-alpha with aggravation of disease. *J Immunol* 169:929-936.

Hoffmann, K.F., Wynn, T.A., Dunne, D.W. (2002) Cytokine-mediated host responses during schistosome infections; walking the fine line between immunological control and immunopathology. *Adv Parasitol* 52:265-307.

Homeida, M., Ahmed, S., Dafalla, A.A., Suliman, S., Eltom, I., Nash, T.E., Bennett, J.L. (1988) Morbidity associated with *Schistosoma mansoni* infection as determined by ultrasound: a study in Gezira, Sudan. *Am J Trop Med Hyg* 39:196-201.

Kabatereine, N.B., Vennervald, B.J., Ouma, J.H., Kemijumbi, J., Butterworth, A.E., Dunne, D.W., Fulford, A.J. (1999) Adult resistance to schistosomiasis mansoni: age-dependence of reinfection remains constant in communities with diverse exposure patterns. *Parasitology* 118: 101-105.

Kamel, I.A., Elwi, A.M., Cheever, A.W., Mosimann, J.E., Danner, R. (1978) *Schistosoma mansoni* and *S. haematobium* infections in Egypt. IV. Hepatic lesions. *Am J Trop Med Hyg* 27:931-938.

Kardoff, R., Stelma, F.F., Vocke, A.K., Yazdanpanah, Y., Thomas, A.K., Mbaye, A., Talla, I., Niang, M., Ehrich, J.H., Doehring, E., Gryseels, B. (1996) Ultrasonography in a Senegalese community recently exposed to *Schistosoma mansoni* infection. *Am J Trop Med Hyg* 54: 586-590.

Kariuki, H.C., Mbugua, G., Magak, P., Bailey, J.A., Muchiri, E.M., Thiongo, F.W., King, C.H., Butterworth, A.E., Ouma, J.H., Blanton, R.E. (2001) Prevalence and familial aggregation of schistosomal liver morbidity in Kenya: evaluation by new ultrasound criteria. *J Infect Dis* 183: 960-966.

Katayama, H., Yokoyama, A., Kohno, N., Sakai, K., Hiwada, K., Yamada, H., Hirai, K. (2002) Production of eosinophilic chemokines by normal pleural mesothelial cells. *Am J Respir Cell Mol Biol* 26:398-403.

Kuhns, D.B., Alvord, W.G., Gallin, J.I. (1995) Increased circulating cytokines, cytokine antagonists, and E-selectin after intravenous administration of endotoxin in humans. *J Infect Dis* 171:145-52.

Lehman, J.S., Jr., Mott, K.E., Morrow, R.H., Jr., Muniz, T.M., Boyer, M.H. (1976) The intensity and effects of infection with Schistosoma mansoni in a rural community in northeast Brazil. *Am J Trop Med Hyg* 25: 285-294.

Lukacs, N. W. (2000) Migration of helper T-lymphocyte subsets into inflamed tissues. *J Allergy Clin Immunol* 106:S264-269.

Mohamed-Ali , Q., Elwali, N.E, Abdelhameed, A.A., Mergani, A., Rahoud, S., Elagib, K.E., Saeed, O.K., Abel, L., Magzoub, M.M., Dessein, A.J (1999) Susceptibility to periportal (Symmers) fibrosis in human *Schistosoma mansoni* infections: evidence that intensity and duration of infection, gender, and inherited factors are critical in disease progression. *J Infect Dis* 180:1298-306.

Mwatha, J.K., Jones, F.M., Mohamed, G., Naus, C.W., Riley, E.M., Butterworth, A.E., Kimani, G., Kariuki, C.H., Ouma, J.H., Koech, D., Dunne, D.W. (2003) Associations between anti-*Schistosoma mansoni* and anti-*Plasmodium falciparum* antibody responses and hepatosplenomegaly, in Kenyan schoolchildren. *J Infect Dis* 187: 1337-41.

Mwatha, J.K., Kimani, G., Kamau T., Mbugua, G.G., Ouma J.H., Mumo, J., Fulford, A.J.C., Jones, F.M., Butterworth, A.E. Roberts, M. & Dunne, D.W. (1998) High levels of TNF, Soluble TNF receptors, sICAM-I and IFN-γ, but low levels of IL-5, are associated with hepatosplenic disease in human schistosomiasis mansoni. *J. Immunol* 160:1992-1999.

Omer, AH., Hamilton PJ, Marshall TF, Draper CC. (1976) Infection with *Schistosoma mansoni* in the Gezaire area of the Sudan. *J Trop Med Hyg* 79:151-157.

Ongom, V.L., Bradley, D.J. (1972) The epidemiology and consequences of *Schistosoma mansoni* infection in West Nile, Uganda. 1. Field studies of a community at Panyagoro. *Trans R Soc Trop Med Hyg* 66:835-851.

Rabello, A. (1995) Acute human schistosomiasis mansoni. *Mem Inst Oswaldo Cruz* 90:277-280.

Richter, J., Hatz, C., Campagne, G., Bergquist, N. R. & Jenkins, J. M. (2000) Ultrasound in schistosomiasis. A practical guide to the standardized use of ultrasonography for the assessment of schistosomiasis related morbidity. Geneva: World Health Organization, TDR/STR/SCH/00.1.

Roux, J.F., Sellin, B., Picq, J.J. (1980) Etude epidemiologique sur les hepato-splenomegalies en zone d'endemie bilharzienne a *Schistosoma mansoni*. *Medicine Tropicale* 40:45-51.

Slungaard, A., Vercellotti, G. M., Walker, G., Nelson, R. D., Jacob, H. S. (1990) Tumor necrosis factor alpha/cachectin stimulates eosinophil oxidant production and toxicity towards human endothelium. *J Exp Med* 171:2025-2041.

Siongok, T.K.A., Mahmoud, A.A.F., Ouma, J.H., Warren, K.S., Muller, A.S., Hander, A.K., Houser, H.B. (1976) Morbidity in schistosomiasis mansoni in relation to intensity of infection: study of a community in Machakos, Kenya. *Am J Trop Med Hyg* 25:273-284.

Smith, D.H., Warren, K.S., Mahmoud, A.A.F. (1979) Morbidity in Schistosomiasis mansoni in relation to intensity of infection: study of a community Kisumu, Kenya. *Am J Trop Med Hyg* 28: 220-229.

Stelma F.F, van der Werf M., Talla I., Niang M., Gryseels B. (1997) Four years' follow-up of hepatosplenic morbidity in a recently emerged focus of *Schistosoma mansoni* in northern Senegal. *Trans R Soc Trop Med Hyg* 91: 29-30.

Symmers, W.S.C. (1904) Note on a new form of liver cirrhosis due to the presence of the ova of *Bilharzia haematobia. J Pathol Bacteriol* 9:237-239.

Vennervald, B.J., Booth, M., Butterworth, A.E., Kariuki, C.H., Kadzo, H., Ireri, E., Amaganga, C., Gachuhi, K., Kenty, L.C., Mwatha, J.K., Ouma, J.H., Dunne, D.W. (In Press a) Regression of hepatosplenomegaly in Kenyan school-aged children after praziquantel treatment and in the absence of re-infection by *Schistosoma mansoni. Trop. Med. Int Health.*

Vennervald, B.J., Kenty, L.C., Butterworth, A.E., Kariuki, C.H., Kadzo, H., Ireri, E., Amaganga, C., Kimani, G., Mwatha, F.K., Otedo, A., Booth, M., Ouma, J.H., Dunne, D.W. (In Press b) Detailed clinical and ultrasound examination of children and adolescents in a *Schistosoma mansoni* endemic area in Kenya: hepatosplenic disease in the absence of portal fibrosis. *Trans R. Soc Trop Med Hyg.*

Walters, J.H. & McGregor, I.A. (1960). The mechanism of malarial hepatomegaly and its relationship to hepatic fibrosis. Trans Roy Soc Trop Med Hyg 54:135-145.

Warren, K.S. (1975) Hepatosplenic schistosomiasis mansoni: an immunologic disease. *Bull N Y Acad Med* 51:545-550

Whittle, H., Gelfand, M., Sampson, E., Purvis, A., Weber, M. (1969) Enlarged livers and spleens in an area endemic for malaria and schistosomiasis. *Trans R Soc Trop Med Hyg* 63:353-361.

Yazdanpanah, Y., Thomas, A.K., Kardorff, R., Talla, I., Sow, S., Niang, M., Stelma, F.F., Decam, C., Rogerie, F., Gryseels, B., Capron. A., Doehring, E., (1997) Organometric investigations of the spleen and liver by ultrasound in Schistosoma mansoni endemic and nonendemic villages in Senegal. *Am J Trop Med Hyg* 57:245-249.

Zhang, H., Singer, B. (1999) *Recursive Partitioning in the Health Sciences.* New York, Springer-Verlag.

Zwingenberger, K., Irschick, E., Vergetti Siqueira, J.G., Correia Dacal, A.R., Feldmeier, H. (1990) Tumour necrosis factor in hepatosplenic schistosomiasis. *Scand J Immunol* 31: 205-211.

Chapter 11

PATHWAYS TO IMPROVED, SUSTAINABLE MORBIDITY CONTROL AND PREVENTION OF SCHISTOSOMIASIS IN THE PEOPLE'S REPUBLIC OF CHINA

Donald P. McManus[1]*, Zheng Feng[2], Jiagang Guo[2], Yuesheng Li [1,3], Paul B. Bartley[1], Alex Loukas[1] and Gail M. Williams[1]

[1]*Australian Centre for International and Tropical Health and Nutrition (ACITHN), The Queensland Institute of Medical Research and The University of Queensland, Brisbane, Australia;* [2]*Institute of Parasitic Diseases, Chinese Center for Disease Control and Prevention, Shanghai, PR China;* [3]*Hunan Institute of Parasitic Diseases, Yueyang, PR China. *Corresponding author; email: donM@qimr.edu.au*

Key words: *Schistosoma japonicum*; morbidity control; China; integrated control program; artemether; immunogenetics of severe disease; transmission blocking vaccine; vaccination; mathematical modelling

1.   INTRODUCTION

600 million Chinese have parasitic infections. One of the most important, from the public health and clinical perspectives, is undoubtedly schistosomiasis (Hotez et al., 1997). Schistosomiasis japonica is a serious disease, typified by long-term chronicity, and a major disease risk for more than 40 million people living in the tropical and sub-tropical zones of China (Chen and Feng, 1999). The disease disabled and killed millions of Chinese peasants before systematic control programs began in the 1950s (Chen, 1999). Infection remains a major public health concern despite the 45 years of extraordinary historic emphasis and intensive efforts at controlling this disease. Even today, it is estimated that over 1 million people and several hundred thousand livestock, including 100,250 bovines, are infected (Ross et al., 2001). The major endemic foci are in the marsh and lake regions (Dongting and Poyang Lakes; Figure 11-1) of Southern China that cover a vast

area of five provinces, Jiangsu, Anhui, Hubei, Jiangxi and Hunan. Since 1985, the rural Chinese economy has been boosted (resulting in an improved standard of living) but the prevalence of *Schistosoma japonicum* and its associated morbidity have also risen (Yuan, 1992). Unlike the other schistosome species known to infect man, the oriental schistosome is zoonotic, with a range of mammalian reservoirs that complicates control efforts. Schistosome infection debilitates domestic livestock that are used for food and as work animals in Southern China. This adds to the economic burden and suffering of communities in the *S. japonicum*-endemic areas.

Clinical features of schistosomiasis japonica are severe, ranging from fever, headache and lethargy, to serious fibro-obstructive pathology leading to portal hypertension, ascites and hepatosplenomegaly that can cause premature death (Ross et al., 2002). As well, infected children are stunted and have cognitive defects impairing memory and learning ability. The onset of advanced schistosomiasis poses a particular public health and socio-economic problem. There are over 30,000 cases of advanced disease with 8,000 new cases reported annually in China (Qing-Wu et al., 2002). The onset of advanced disease promotes the most serious clinical manifestations and it is difficult to provide adequate treatment for affected patients, many of whom die resulting in severe consequences both for the patient's family and the community as a whole. As well, P53 gene mutations have been reported in rectal cancer associated with advanced schistosomiasis japonica in Chinese patients (Zhang et al., 1998).

2. MORBIDITY CONTROL OF SCHISTOSOMIASIS JAPONICA IN CHINA

Current control in China is based on community chemotherapy with a single dose of praziquantel. Vaccines (for use in bovine and humans) in combination with other control strategies, including the use of new drugs, are needed to make elimination of the disease possible (McManus and Bartley, in press). New approaches, advocated by the Chinese Government and other organizations including WHO, are required to complement current control strategies that are difficult and costly (Guyatt, 1998, 2003). We recently brought together the results of control technology advances, including the success of recent World Bank Loan Project (WBLP) (1992-2000) inputs to mathematically model prospects for the future control of schistosomiasis in China (Williams et al., 2002). The model (see below) shows that an integrated approach to control can lead to possible elimination of infection. Whereas appropriate human drug treatment declines worm burden and the intensity of infection as well as prevalence, chemotherapy alone cannot reduce transmission substantially because of the zoonotic nature of the disease.

China is in the midst of major environmental and social change and a further important issue concerns construction of the giant Three Gorges Dam (Figure 11-1) across the Yangtze (Chang Jiang) River, whose drainage provides the great corridor of parasitic disease in China (Davis et al., 1999). Due to be filled by 2009, the dam, which was recently closed, and the 600 km long reservoir it will create, will have a global environmental impact on the transmission and geographical distribution of schistosomiasis both above and below the dam (Zheng et al., 2002). The Ministry of Health in China is currently investigating the increased risk for schistosomiasis associated with the dam. No absolute schistosomiasis data exist to indicate the probable outcomes but there is no doubt the dam will change the Yangtze basin ecology significantly. A second wave of problems has been caused by a plague of recent floods that have devastated numerous regions of the Yangtze River drainage (Davis et al., 1999). The recent flood problems are associated with increasing deforestation in the mountains coupled with reclaiming of lakes and marshy lowlands in the Lake Districts in the lower Yangtze drainage (Davis et al., 1999). This will increase snail habitats and, consequently, the risk of schistosome transmission. Over the coming years the numbers of oncomelanid-snail breeding areas will be increased substantially and schistosomiasis transmission will intensify. As a result we predict a marked elevation in infection rates and associated morbidity that will impact significantly on control, providing new challenges for the national control program in China.

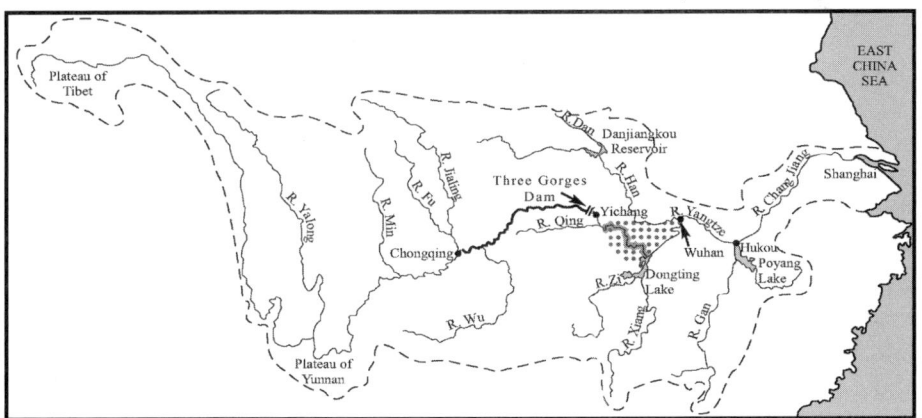

Figure 11-1. Map of Southern China showing the position of the Three Gorges Dam and Poyang and Dongting Lakes on the River Yangtze.

A recent volume of *Acta Tropica* (volume 82, 2002; 93-319) was devoted entirely to a number of important papers that are directly relevant to this article. The papers were presented at an International Symposium on Schistosomiasis, sponsored by the Ministry of Health, PR China (July, 2001), with the theme of "Schistosomiasis control in the 21st century" and emphasized studies on Asian schistosomiasis.

3.   SOME RESEARCH QUESTIONS THAT NEED TO BE ADDRESSED

We are involved in a National Institutes of Health-supported Tropical Medicine Research Center (TMRC) that seeks to address the issue of "Emerging Helminthiases in China: Genetic Diversity, Transmission Dynamics, and the Impact of Environmental Change". Focused through the Institute of Parasitic Diseases, Chinese Centre for Disease Control and Prevention, the TMRC project comprises four research components: a). *Schistosoma japonicum*: genomics, post-genomics and genetic diversity; b). Ecogenetics of *Schistosoma japonicum* transmission in China; c). Environmental change impacting on control and transmission of *Schistosoma japonicum*; and d). Population genetics and transmission patterns of hookworms in China. A strong collaboration between China-, USA- and Australia-based scientists cemented in the TMRC forms the basis of a very strong scientific milieu for the undertaking of the research.

Answers to additional research questions are required for an enhanced appreciation of the far-reaching importance of the spread and transmission dynamics of Asian schistosomiasis in China that will result in improved strategies for control. Furthermore, research is required that will investigate the immunogenetics of severe *S. japonicum* infection, and that expands into genomics and post-genomics research for the development of new diagnostics and tools for interventions such as vaccines and new drugs. Mathematical modeling is required that can predict the impact on schistosome disease morbidity and the cost effectiveness of schistosomiasis control strategies that will be affected by the use of the new interventions and by the changing ecology that will occur as result of the construction of the 3 Gorges Dam. Much of this future research will be focused on the Poyang (Jiangxi Province) and Dongting (Hunan Province) lakes, the two largest lakes in China, and Hubei Province, where the majority of schistosomiasis japonica cases occur (Figure 11-1).

3.1 Laboratory, field and hospital studies of the effects of artemether on *S. japonicum* infection

Praziquantel (pzq) is highly effective in decreasing the morbidity of schistosomiasis, but fails to prevent re-infection (Li et al, 2002). In addition, pzq cannot be used preventatively as it is effective only on adult worms and very early stage schistosomula. There is not yet clear-cut evidence for the existence of pzq-resistant schistosome strains, but there has been considerable discussion whether or not there is resistance, and decreased susceptibility to the drug has been observed in several countries (Doenhoff et al., 2002). There is no direct proof of resistance to pzq in *S. japonicum* (Yu et al., 2001). Nevertheless, there is a need for research and development of novel and inexpensive drugs against schistosomiasis, especially schistosomiasis japonica, as no viable alternative exists (WHO, 2002). Artemether, a derivative of artemesinin (qinghaosu), has been used extensively in treatment of malaria, but it also shows an effect on schistosomes especially on 5-21-day-old worms as demonstrated by the pioneering studies of Xiao Shuhua and his collaborators (Xiao et al., 2002). Artemether and pzq thus act against different development stages of schistosomes. There is a need to establish a position for artemether in the control of schistosomiasis in China by assessing its role in (a). prophylactic effects against acute *S. japonicum* infection, and (b). combination treatment with praziquantel for acute schistosomiasis japonica.

It has been hypothesized that combined treatment would result in higher worm reduction rates than a single dose of pzq. This hypothesis has been confirmed experimentally with rabbits and hamsters (Xiao et al, 2000, 2002) but, to date, praziquantel and artemether combinations have not undergone clinical trials (Utzinger et al., 2003). However, a combination of pzq and artesunate (another derivative of artemesinin) was used in a non-blinded open-label treatment trial with *S. mansoni* in Senegal (De Clercq et al., 2000) and a randomized controlled clinical trial with *S. haematobium* in Gabon (Borrman et al., 2001) with encouragingly beneficial outcomes. If a successful clinical outcome occurs with the pzq/arthemether combination, this will serve as the basis for combination chemotherapy resulting in improved control in specific endemic settings.

In brief, we plan to: (i). Further document the prophylactic effects of artemether against incident acute *S. japonicum* infections in various transmission zones in Hubei and Hunan provinces, and (ii). Monitor the efficacy of combined treatment with pzq and artemether on acute cases of *S. japonicum* infection in local hospitals. The specific objectives are: 1). To document the prophylactic effects of artemether against acute schistosome infections in the transmission seasons; 2). To validate benefits of combination artemether and pzq treatment of acute schistosomiasis at the hospital level. 3). To examine the effects of artemether treatment on human immunity to re-infection. This latter study will determine whether prophylactic treatment with artemether confers immunity to re-infection in humans. It will also investigate immunological correlates associated with any protective immunity that is generated. Resistance following attenuation of early *Schistosoma mansoni* infections in mice has been reported with another drug (Ro 11-3128) (Bickle and Andrews, 1985) although the immunity induced is species-specific and does not operate against *S. japonicum* (Bickle et al., 1990).

Artemether interacts with haeme to generate an organic free radical that alkylates and denatures protein - resulting in the death of 5-21-day old schistosomules (Xiao et al., 2002). The schistosomules die at a similar time as radiation-attenuated cercariae (the current "gold-standard" vaccine). Artemether-affected schistosomules are infiltrated with host lymphocytes - allowing exposure to the immune system of previously hidden molecules. We hypothesize that artemether treatment in this time-period provides protection similar to that offered by irradiated cercariae - providing a "gold-standard" model that can be used to investigate experimentally the mechanisms of protective immunity with practical implications for the management of human disease through vaccination.

3.2   Immunogenetics of severe *Schistosoma japonicum* infection in China

As already emphasized, praziquantel chemotherapy has been the cornerstone of control activities that target infected individuals but it has been shown that pzq treatment has little effect on existing fibrotic lesions (Li et al., 2000). It is important to determine whether there is any genetic component associated with human susceptibility to advanced schistosomiasis japonica as has been shown for schistosomiasis mansoni (Dessein et al., 1999). If this is the case, it should be possible to identify those Chinese individuals susceptible to late stage fibrotic disease and provide appropriate treatment. The principal aim is to identify allelic variants that increase human susceptibility to advanced fibrosis in *S. japonicum* infections. Immune correlates associated with disease susceptibility (Henri et al., 2002) can then be additionally sought.

Various HLA class II associations with severe disease in *S. japonicum* (Waine et al., 1998; Hirayama et al., 1998, 1999; McManus et al., 2001) have been reported. There has been one published study (Liu et al., 1999) on a small number of advanced cases that suggested family aggregation of advanced schistosomiasis japonica does exist. What is now required is an extension of this work to involve identification of sibling (sib)-pairs with advanced disease, then to search for a genetic locus in the sib-pairs associated with disease susceptibility and, subsequently, for susceptibility genes in the identified locus. A logical research design to follow is that formulated by Dr Alain Dessein (Institut national de la Sante et de la Recherche Medicale (INSERM), Faculty of Medecine, Marseille France) who examined the causes of advanced fibrosis in *S. mansoni* infected subjects living in two Sudanese populations. Advanced fibrosis was aggregated in certain families suggesting that such factors could be genetic. Segregation analysis of the phenotype provided evidence for the major SM2 locus control of fibrosis progression from mild to severe fibrosis associated with portal hypertension (Dessein et al., 1999). This locus SM2 was mapped by linkage analysis in the region 6q23. Further, SM2 was located close to IFN- receptor 1 that encodes the alpha chain of the IFN- receptor. One interpretation of these data is that mutations in this receptor lead to loss of receptor function that, in turn, is associated with a lack of effectiveness of IFN- in suppressing fibrogenesis.

In order to meet the desired objectives it will be necessary to:

1). Undertake a hospital-based survey of advanced schistosomiasis cases in communities in China with a long history of human schistosomiasis endemicity. The defined target phenotype is fibrosis characterized by the presence of abdominal ascites fluid. The degree of peri-portal fibrosis and patterns of parenchyma can be estimated according to the classification proposed by the World Health Organization Cairo Working Group (Hatz, 2001) and a recent meeting on ultrasound criteria for mekongi and japonicum schistosomiasis (coordinator, Dr Christoph Hatz, Swiss Tropical Institute) of the Regional Network for Asian Schistosomiasis, Phnom Penh, Cambodia, May 27-29, 2002).

2). Trace the nuclear family of individual cases and identify sib-pairs that are afflicted with advanced disease.

3). Identify genetic regions carrying gene(s) implicated in the control of advanced fibrosis. 4). Test whether the same loci are involved in *S. mansoni* and *S. japonicum* infections.

5). Identify in these genetic regions, genetic polymorphisms that are linked to the control of advanced fibrosis.

6). Compare the production of pro- and anti-fibrogenic cytokines in subjects with chronic and advanced schistosomiasis to determine whether specific cytokines correlate with the development of advanced disease.

A better knowledge of the pathogenesis of advanced fibrosis in *S. japonicum* infections and the identification of the susceptibility genes will allow investigation of new therapeutic avenues, preventive treatments (including vaccination) as well as curative therapies.

3.3 Bovine intervention trial using a cluster-randomized design

A TMRC-supported pilot intervention study concluded in late 2002 around the Poyang Lake, Jiangxi Province aimed at providing proof-of principle for the hypothesis that bovine, especially buffalo, infections are responsible for the persistence of human schistosomiasis transmission in the marshlands and lake areas of China. Underpinning this work was the fact that despite numerous descriptions of the importance of bovines in schistosomiasis transmission in the literature, no group had hitherto attempted to scientifically prove the basic hypothesis. The work has major implications for future integrated schistosomiasis control including the use of chemotherapy for bovines and it underpins the rationale for development and implementation of a veterinary-based transmission-blocking vaccine for use in reservoir hosts, particularly buffaloes.

The pilot study (1998-2002) compared human infection incidence in an intervention village (Jishan) (in which buffaloes were treated) and a control village (Hexi) (no buffalo treatment) (Guo et al., 2001; Davis et al., 2002). On completion of the pilot study, the prevalence of infection in both residents and buffaloes were reduced significantly in the intervention village. The intervention village experienced a highly significant 70% reduction in human infection prevalence, compared to the non-intervention village. This effect became apparent at the end of three years of intervention, and was confirmed at the end of the fourth year (2002). Detailed adjustment for water contact (including specifically snail-inhabited water) over the study period confirmed the findings of the effectiveness of bovine treatment at the level of 70%. Based on data collected in Jishan village, the predicted human infection prevalence was used in the mathematical model described by Williams et al. (2002) to compare predicted prevented infections associated with bovine treatment (Figure 11-2).

This was a promising pilot study that provides proof of principle but it is important to confirm the results in a larger, more generalizable cluster-randomized design, across the full range of transmission modes, to accurately assess efficaciousness. This new study will involve selection of 6 pairs of villages from 24 identified communities (human prevalence 10-15%; bovine prevalence 10-26%) in the Lakes area of China with pairs being matched as closely as possible on factors related to transmission: i.e. infection levels, transmission ecology (e.g.: water height, flooding patterns), and force of infection (e.g.: water contact patterns, buffalo:human numbers, sanitation practices, herding practices). The trial will initially target around 16 villages, and aim to select the 12 that provide closest matching in pairs. Within each pair, one village will be randomly chosen as intervention (buffaloes treated annually) and the other will be a control. Annual mass treatment of residents will be carried out in both villages. The end-point for the trial will be human infection rates. The efficacy of buffalo treatment will be measured by the accumulated comparison of human infection rates across intervention and control villages, and will be reflected in the predicted percentage of human infection-years prevented by the intervention. Human infection rates will be assessed at the end of each year for 4 years. This will be done by randomly selecting an age-sex stratified sample of 200 residents per village, per year.

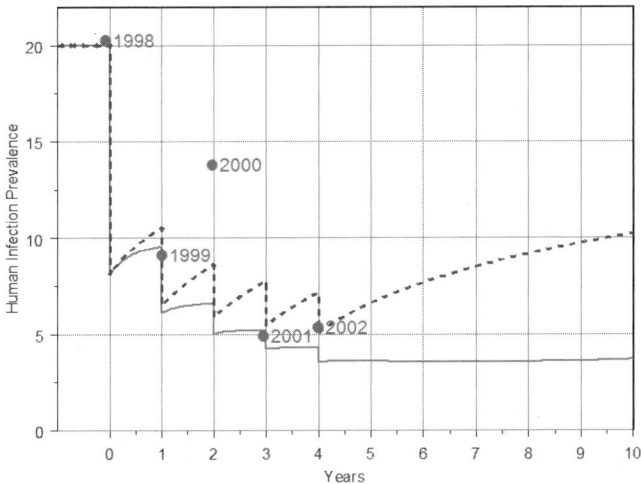

Figure 11-2. Predicted human infection prevalence from the mathematical model of Williams et al. (2002), based on data collected in Jishan village, for human treatment only (broken line) and human and bovine treatment combined (solid line). Solid dots show the prevalences actually observed in Jishan. The gap between the two curves represents predicted prevented infections associated with bovine treatment.

Given that it is expected (based on results of the proof-of-principle trial) that a reduction of 50% will occur, and based on the an average infection rate of 15% in non-intervention villages, and a design effect (relating to paired differences) of 3, the trial will have at least 80% power to detect the intervention effect. This is a conservative calculation, since the accumulation of results over 4 years will increase power, although the (expected) decline in prevalence (associated with human treatment) will limit the increase. An effectiveness measure will be estimated, with confidence intervals taking into account the observed design effect. Supplementary analysis will also examine treatment effectiveness (as measured by human and bovine cure rates), and human water contact and sanitation patterns for each village in the study. The use of a larger number of villages will add the important component of generalizability, as well as power to the findings. The data from this trial will also inform the modeling studies, outlined below, since they will contribute to empirical validation of the model.

3.4   Mathematical modeling

Building on results from previous, ongoing and planned epidemiological studies of schistosomiasis japonica in China, and the associated development and validation of the published mathematical model for schistosomiasis japonica (Williams et al., 2002) (The basic model is shown in Figure 11-3), it is now timely to: compare the impact of competing strategies (human chemotherapy options, mass bovine chemotherapy, bovine vaccines, chemo-prophylaxis, improvements in human sanitation, reduction in water contact) and combinations of strategies on the burden of chronic and advanced schistosomiasis japonica infection in a variety of transmission ecologies; compare the impact of competing strategies on the development of new cases; compare the cost-effectiveness of competing strategies over 5, 10, and 20 year time frames; and undertake quantitative comparison of factors influencing different transmission modes in China, including the Three Gorges Dam.

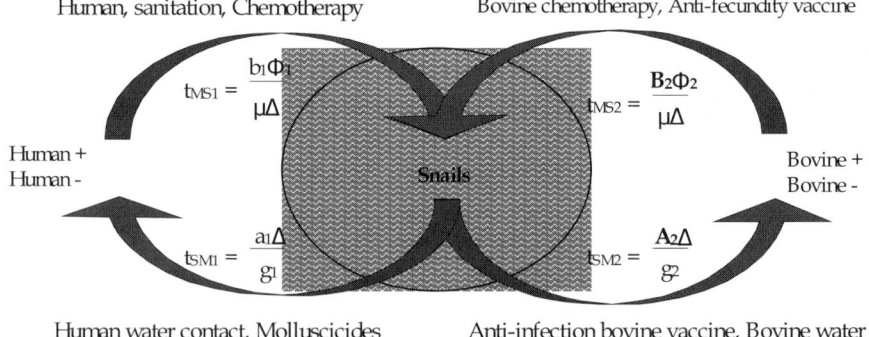

Figure 11-3. Mathematical model for S. japonicum transmission, showing parameters and intervention pathways (After Williams et al., 2002).

The model utilizes a set of simultaneous differential equations that mimic the transmission process. Endemic infection rates in humans, bovines and snails plus relevant parameters enable the prediction of prevalence over time.   Control scenarios include human and bovine chemotherapy (mass, screening or targeted, with variable efficacy, coverage and periodicity), anti-infection and anti-fecundity bovine vaccines (variable efficacy and coverage), artemether, reduction in human and bovine water contact, human sanitation, and reduction in bovine populations from the transmission site. Exacerbation scenarios are also envisaged, notably the environmental impact of the Three Gorges Dam. Preliminary predictions of this include an increase in snail habitat and survival (current survival of snails in endemic areas is curtailed by flooding, which will be more controlled when the Three Gorges Dam is completed). The model allows for heterogeneity among human and bovine hosts, and for parameters to be adjusted to fit a particular transmission ecology. The model parameters are altered to reflect a specific control/exacerbation scenario, and model equations used to predict prevalence of infection over time. Two overall measures of effectiveness at a given time are available: infected person years prevented, and new cases prevented. Data relating to the relationship between infection and morbidity will be reviewed, so that predictions of morbidity rates can be made from predicted prevalence and incidence of infection.

Effectiveness as defined above will be assessed at 5, 10 and 15 years from initiation of the intervention program. A sub-component of the World Bank Loan Project for Schistosomiasis Control (WBLPSC) (Yuan et al., 2000), aimed to carry out a cost-effectiveness and cost-benefit analysis. Based on information available from WBLPSC completion reports, however, this appears to have only utilized operational costings, such as cost of case detection and treatment, cost of reducing snail infested areas and cost of reduction of bovine infections. This is insufficient to enable control scenarios to be compared for cost-effectiveness. A cataloguing of the information needed for costing of various control strategies has been done, resulting in a cost menu for each control strategy.

Newly obtained epidemiological data can be used to validate the mathematical model further. Validation consists of using parameters estimated from data obtained from population and cohort studies (endemic prevalence, treatment efficacy and coverage, changes in water contact, for example), and using the model to obtain predicted prevalences that are compared to observed prevalences. Once validated, predictions of longer-term effects from control strategies, alone or in combination, and the comparative cost-effectiveness can be assessed at the appropriate intervals of time. The value of the modeling lies in the ability to make longer-term predictions, based on optimized strategies, or their combination, than would be possible to test directly in field-based experiments. By allowing the parameter values to vary over a distribution, the values needed to achieve a given effectiveness or cost-effectiveness can be identified. This will facilitate the setting of targets for program achievement, such as coverage and efficacy. The predicted outcomes of this work are: the further development of a validated mathematical model for schistosomiasis japonica that will inform the optimal choice of disease control programs (including combinations of interventions), in terms of effectiveness and cost; demonstration of the relative magnitudes of short and long-term benefits of control strategies; setting of targets for intervention development (treatment coverage, vaccine efficacy, health promotion uptake) that are needed to achieve targets for disease reduction; and the customization of schistosomiasis control strategies to specific transmission ecologies.

3.5   Post-genomics research leading to new vaccines and diagnostics for public health use

A highly adapted relationship exists between schistosomes and their hosts that appears to involve parasite exploitation of host endocrine and immune signals. Aspects of parasite biology, drug resistance, and immune evasion strategies that determine avoidance of the host immune system have long perplexed clinicians and investigators intent on controlling this parasitic disease. An expanded knowledge of the schistosome genome has become increasingly important for understanding these complex issues as well as other biomedical aspects of schistosomiasis, including new insights into invertebrate evolution. Consequently, a major gene discovery program for *S. japonicum* is currently underway in Shanghai focused through the Chinese National Human Genome Center (CHGC) and the Institute of Parasitic Diseases, Chinese Center for Disease Control and Prevention. Transcriptome information, obtained from ~50000 randomly selected cDNA clones from female, male, mixed sex groups of adult worms, and *S. japonicum* eggs, was recently reported (Hu et al., in press). The 13131 gene clusters represented by these expressed sequence tags (ESTs) likely encode the majority of the protein-encoding genes of schistosomes given that these parasites may have ~15,000 - 20000 genes (Johnston et al., 1999).

The EST data have shown that, remarkably, *S. japonicum* encodes receptors for mammalian insulin, progesterone, cytokines and neuropetides, suggesting that host hormones orchestrate schistosome development and maturation, and inferring that schistosomes moderate anti-parasite immune responses through inhibitors, molecular mimicry and other strategies. The availability of these and the other EST sequences should lead to a more profound understanding of the schistosome genome, molecular pathogenesis of schistosomiasis, and the development of new intervention strategies. Furthermore, the availability of such a large number of new *S. japonicum* ESTs will allow studies on identification of *S. japonicum* stage-specific gene transcripts by cDNA microarray profiling (Hoffmann et al., 2002).

A new genomic sequencing project is underway at CHGC that aims to sequence the entire genome of *S. japonicum*. A BAC library is under construction and 2.1 X shotgun sequencing coverage has been completed with 6 X coverage planned. This will allow the undertaking of wide scale post-genomics research focusing on the newly discovered genes, especially characterization of candidate molecules, such as secretory/excretory and surface proteins, as intervention and diagnostic targets (Smyth et al., 2003).

Of several molecules that we are especially targeting for study, one is paramyosin (pmy; Sj-97), a myofibrillar protein found exclusively in invertebrates and a primary vaccine candidate antigen against schistosomiasis. The results of vaccine trials against *S. japonicum* undertaken on inbred and outbred mice and water buffaloes using a bacterially expressed and purified form of *S. japonicum* pmy (rec-Sj-97) have been reported (McManus et al., 2001). Vaccination of the mice resulted in high levels of specific anti-pmy IgG antibodies when compared with quil A adjuvant controls and significant reduction in worm burdens and in liver eggs. Furthermore, a significant reduction in liver eggs was recorded in two of the three water buffalo vaccine trials undertaken and, in all three trials, high levels of specific anti-pmy IgG antibodies were generated. The development of a vaccine intended for livestock animals such as bovines would be beneficial in two ways; directly by blocking transmission of schistosomiasis to humans and economically by contributing to healthier livestock. We are encouraged by the consistent efficacy in the mouse and the buffalo vaccine trials that resulted in a significant decrease in liver eggs. Indeed, predictions from the mathematical model (Williams et al., 2002) indicates that an egg reduction effect of 42-45% in buffaloes, which we have consistently obtained, would be sufficient when combined with human treatment plus other existing control strategies to eliminate schistosomiasis in the marshland and lake regions along the middle and upper reaches of the Yangtze River. This is illustrated in Figure 11-4.

Routine expression of the Sj-97 protein is as a PQE30 construct in *E. coli* with subsequent purification on metal affinity columns. Whereas the yields of rec-paramyosin are sufficient for small trials, the expression levels are low and are inadequate for large-scale use. Accordingly, codon-optimized expression of Sj-97 in yeast is now underway in collaboration with Dr WeiQing Pan (Second Military Medical University, Shanghai) who has had considerable success in expressing large malaria proteins in a variety of hosts in a synthetic gene format that can prevent degradation of proteins (Pan et al., 1999). The yeast-expressed Sj-97 will be tested in mice in Brisbane and then in buffalo field trials in China, using quil A adjuvant and vaccine/challenge procedures that we have used with success previously (McManus et al., 2001).

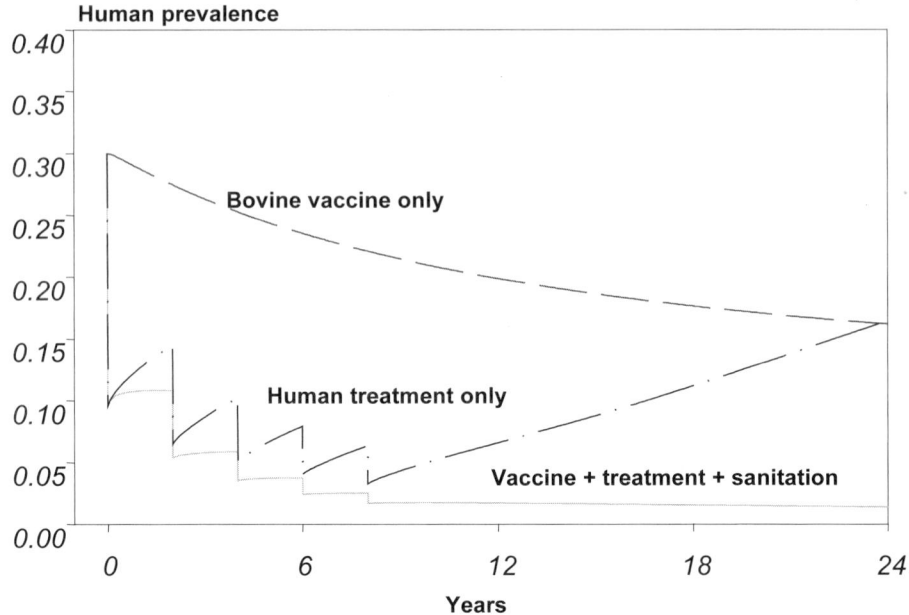

Figure 11-4. Strategies for schistosomiasis control: the predicted effect on human prevalence of the bovine vaccine and human praziquantel treatment alone or in combination. Note that whereas the vaccine or treatment alone have limited impact over time, a combination of control measures can lead to very low prevalence approaching elimination.

4.   CONCLUDING COMMENTS

Control of schistosomiasis in China is based on praziquantel treatment. Vaccines and other control methods, including new drugs, are needed to make disease elimination possible. Control will be complicated by construction of the Three Gorges Dam across the Yangtze River. This will increase snail breeding areas and markedly elevate infection rates and disease. Several areas have been highlighted for future study that will: determine whether another drug, artemether, can be used as an aid in control; increase our understanding of the immunological and genetic processes involving in the development of disease; ratify the importance of buffalo infections in maintaining human transmission; undertake genomics and post-genomics research on existing and newly discovered *S. japonicum* molecules that are vaccine candidates diagnostics; and develop a mathematical model that can predict the optimum methods for the sustained control of schistosomiasis japonica that will be of significant value to the public health of China and the Chinese people.

ACKNOWLEDGEMENTS

The authors acknowledge The Wellcome Trust, The National Health and Medical Research Council of Australia, The UNDP/World Bank/WHO Special Programme for Research and Training in Tropical Diseases (TDR), The Chinese High-Tech Research and Development Program and NIAID/NIH, USA (2 P50 AI39461-06A1) for financial support. We also thank our numerous collaborators: Mary Duke, Paul Brindley, George Davis, Adrian Sleigh, Chuck Shoemaker, Christophe Chevillard, Alain Dessein, Honggen Chen, Weiqing Pan, Hu Wei, Ze-Guang Han, Sheng-Yue Wang, Geoff Gobert, Luke Moertel, Malcolm Jones, and Li Hua Zhang.

REFERENCES

Bickle QD and Andrews BJ. Resistance following drug attenuation (Ro 11-3128 or oxamniquine) of early *Schistosoma mansoni* infections in mice. *Parasitology* 1985; 90: 325-38.

Bickle QD, Sacko M, Vignali DA. Induction of immunity against *Schistosoma mansoni* by drug (Ro11-3128)- terminated infections: analysis of surface antigen recognition. *Parasite Immunol.* 1990; 12:569-86.

Borrmann S, Szlezak N, Faucher JF, Matsiegui PB, Neubauer R, Binder RK, Lell B, Kremsner PG. Artesunate and praziquantel for the treatment of *Schistosoma haematobium* infections: a double-blind, randomized, placebo-controlled study. *J. Infect Dis.* 2001; 184:1363-6.

Chen MG. Progress in schistosomiasis control in China. *Chinese Med. J.* 1999;112: 930-933.

Chen MG, Feng Z. Schistosomiasis control in China. *Parasitol. Int..* 1999; 48:11-19.

Davis, GM, Wu, W-P, Chen, H-G, Liu, H-Y, Guo, J-G, Lin, D-D, Lu, S-B, Williams, GM, Sleigh, AC, Feng, Z and McManus, DP. A baseline study of the importance of bovines for human *Schistosoma japonicum* infections around Poyang Lake, China: villages studied and snail sampling strategy. *Am. J. Trop.Med.Hyg.* 2002; 66: 359-71.

Davis, GM, Wilke T, Zhang Y, Xu, XJ, Qiu CP, Spolsky C, Qiu DC, Li Y, Xia MY and Feng Z. Snail-*Schistosoma, Paragonimus* interactions in China: population ecology, genetic diversity, coevolution and emerging diseases. *Malacologia* 1999; 41: 355-77.

De Clercq D, Vercruysse J, Verle P, Kongs A, Diop M. What is the effect of combining artesunate and praziquantel in the treatment of *Schistosoma mansoni* infections? *Trop. Med. Int. Hlth.* 2000; 5: 744-6.

Dessein, AJ, Hillaire, D, Elwali, NE, Marquet, S, Mohamed-Ali, Q, Mirghani, A, Henri, S, Abdelhameed, AA, Saeed, OK, Magzoub, MM and Abel, L. Severe hepatic fibrosis in *Schistosoma mansoni* infection is controlled by a major locus that is closely linked to the interferon-gamma receptor gene. *Am. J. Hum. Genet.* 1999; 65: 709-21.

Doenhoff, MJ, Kusel, JR, Coles, GC and Cioli, D. Resistance of *Schistosoma mansoni* to praziquantel: is there a problem? *Trans. R. Soc. Trop. Med. Hyg.* 2002; 96: 465-9.

Guo, JG, Ross, AGP, Lin, DD, Williams, GM, Chen, HG, Li, YS, Davis, GM, Feng, Z, McManus, DP and Sleigh, AC. A baseline study on the importance of bovines for human *Schistosoma japonicum* infection around Poyang Lake, China. *Am. J. Trop. Med.Hyg.* 2001; 65: 272-78.

Guyatt H. Different approaches to modelling the cost-effectiveness of schistosomiasis control.

Mem Inst Oswaldo Cruz. 1998;93 Suppl 1:75-84.

Guyatt H. The cost of delivering and sustaining a control programme for schistosomiasis and soil-transmitted helminthiasis. *Acta Trop.* 2003; 86:267-74.

Hatz, CF. The use of ultrasound in schistosomiasis. *Adv. Parasitol.* 2001; 48: 225-84.

Henri S, Chevillard C, Mergani A, Paris P, Gaudart J, Camilla C, Dessein H, Montero F, Elwali NE, Saeed OK, Magzoub M, Dessein AJ. Cytokine regulation of periportal fibrosis in humans infected with *Schistosoma mansoni*: IFN-gamma is associated with protection against fibrosis and TNF-alpha with aggravation of disease. *J. Immunol.* 2002;169:929-36.

Hirayama, K, Chen, H, Kikuchi, M, Yin, T, Gu, X, Liu, J, Zhang, S and Yuan, H. HLA-DR-DQ alleles and HLA-DP alleles are independently associated with susceptibility to different stages of post-schistosomal hepatic fibrosis in the Chinese population. *Tissue Antigens* 1999; 53: 269-74.

Hirayama, K, Chen, H, Kikuchi, M, Yin, T, Itoh, M, Gu, X, Zhang, S and Yuan, H. Glycine-valine dimorphism at the 86th amino acid of HLA-DRB1 influenced the prognosis of postschistosomal hepatic fibrosis. *J. Inf.Dis.* 1998; 77: 1682-6.

Hoffmann, KF, Johnston DA and Dunne, DW. Identification of *Schistosoma mansoni* gender-associated gene transcripts by cDNA microarray profiling. Genome Biol. 2002; 3: 1-12.

Hotez, PJ, Zheng, F, Long-qi, X, Ming-gang, C, Shu-hua, X, Shu-xian, L, Blair, D, McManus, DP and Davis, GM. Emerging and reemerging helminthiases and the public health of China. *Emerg. Infect. Dis.* 1997; 3: 303-10.

Hu, W, Yan, Q, Shen, D-K, Liu, F, Zhu, Z-D, Song, H-D, Wang, Z-J, Rong, Y-P, Zeng, L-C, Wu, J, Zhang, X, Wang, J-J, Xu, X-N, Wang, S-Y, Fu, G, Xu, X-R, Zhang, X-L, Wang, Z-Q, Brindley, PJ, McManus, DP, Xue, C-L, Zheng, F, Chen, Z and Han, Z-G. Evolutionary and biomedical implications of a *Schistosoma japonicum* complementary DNA resource. *Nat. Genet.* 2003; 35: 139-47.

Johnston, DA, Blaxter, ML, Degrave, WM, Foster, J, Ivens, AC and Melville, SE. Genomics and the biology of parasites. *BioEssays* 1999; 21: 131-47.

Li, Y, Yu, DB, Li, YS, Ross, AG and McManus, DP. Infections with hepatitis B virus in three villages endemic for schistosomiasis japonica in the Dongting Lake region of China. *Ann. Trop. Med. Parasitol.* 1997; 91: 323-7.

Li, YS, Sleigh, AC, Ross, AG, Li, Y, Williams, GM, Tanner, M and McManus, DP. Two-year impact of praziquantel treatment for *Schistosoma japonicum* infection in China: re-infection, subclinical disease and fibrosis marker measurements. *Trans.Roy. Soc. Trop. Med. Hyg.* 2000; 94: 191-7.

Li, YS, Sleigh, AC, Tanner, M, Dessein, A, Li, Y, Williams, GM and McManus, DP. Five-year impact of repeated praziquantel chemotherapy on sub-clinical morbidity due to *Schistosoma japonicum* in China. *Trans.Roy. Soc. Trop. Med. Hyg.* 2002; 96; 438-43.

Liu Y, Yuan H, Lin D, Hu F, Liu Y, Zhang S, Zhao G, Jiang Q. [Study on family aggregation of cases of advanced schistosomiasis japonica]. [Article in Chinese]. *Zhongguo Ji Sheng Chong Xue Yu Ji Sheng Chong Bing Za Zhi.* 1999;17:149-51.

McManus DP and Bartley PB. A vaccine against Asian schistsomiasis. *Parasitol. Int.* (in press).

McManus, DP, Ross, AGP, Wiliams, GM, Sleigh, AC, Wiest, P, Erlich, H, Trachtenburg, E, Guanling, W, McGarvey, ST, Li, YS and Waine, GJ. HLA class II antigens positively and negatively associated with hepatosplenic schistosomiasis in a Chinese population. *Int. J. Parasitol.*. 2001; 31: 674-80.

McManus, DP, Wong, JYM, Zhou, J, Cai, C, Zeng, Q, Smyth, D, Li, YS, Kalinna, BH, Duke, MJ and Yi, X. Recombinant paramyosin (rec-Sj-97) tested for immunogenicity and vaccine efficacy against *Schistosoma japonicum* in mice and water buffaloes. *Vaccine*

2002; 20: 870-8.

Pan W, Ravot E, Tolle R, Frank R, Mosbach R, Turbachova I, Bujard H. Vaccine candidate MSP-1 from *Plasmodium falciparum*: a redesigned 4917 bp polynucleotide enables synthesis and isolation of full-length protein from *Escherichia coli* and mammalian cells. *Nucleic Acids Res.* 1999; 27: 1094-1103.

Qing-Wu J, Li-Ying W, Jia-Gang G, Ming-Gang C, Xiao-Nong Z, Engels D. Morbidity control of schistosomiasis in China. *Acta Trop.* 2002;82:115-25.

Ross, AGP, Bartley, PB, Sleigh, AC, Olds, GR, Li, YS, Williams, GM and McManus, DP. Schistosomiasis. *New Engl. J. Med.* 2002; 346: 1212-20.

Ross, AGP, Sleigh, AC, Li, Y, Davis, GM, Williams, GM, Jiang, Z, Feng, Z and McManus, DP. Schistosomiasis in the People's Republic of China: prospects and challenges for the 21st century. *Clin. Microbiol Rev.* 2001; 14: 270-95.

Smyth, DJ, McManus, DP, Smout, MJ, Laha, T and Loukas, A. Isolation of cDNAs encoding secreted and transmembrane proteins from *Schistosoma mansoni* using signal sequence trap. *Infect.Immunity* 2003; 71: 2548-54.

Utzinger J, Keiser J, Shuhua X, Tanner M, Singer BH. Combination chemotherapy of schistosomiasis in laboratory studies and clinical trials. *Antimicrob. Agents Chemother.* 2003; 47:1487-95.

Waine, GJ, Ross, AG, Williams, GM, Sleigh, AC and McManus, DP. HLA class II antigens are associated with resistance or susceptibility to hepatosplenic disease in a Chinese population infected with *Schistosoma japonicum*. *Int. J. Parasitol.* 1998; 28: 537-42.

WHO Technical Report Series 912. Prevention and control of schistosomiasis and soil-transmitted helminthiasis. World Health Organisation, Geneva.

Williams, GM, Sleigh, AC, Li, YS, Feng, Z, Davis, GM, Chen, H, Ross, AGP, Bergquist, R and McManus, DP. Mathematical modelling of *Schistosoma japonica*: comparison of control strategies in the People's Republic of China. *Acta Trop.* 2002; 82: 253-62.

Xiao, SH, Booth, M and Tanner, M. The prophylactic effects of artemether against *Schistosoma japonicum* infections. *Parasitol. Today* 2000; 3: 122-6.

Xiao S, Tanner M, N'Goran EK, Utzinger J, Chollet J, Bergquist R, Chen M, Zheng J. Recent investigations of artemether, a novel agent for the prevention of schistosomiasis japonica, mansoni and haematobia. *Acta Trop.* 2002; 82:175-81.

Yuan H, Jiagang G, Bergquist R, Tanner M, Xianyi C, Huanzeng W. The 1992-1999 World Bank Schistosomiasis Research Initiative in China: outcome and perspectives. *Parasitol. Int.* 2000; 49:195-207.

Yuan HC Epidemiological features and control of schistosomiasis japonica in China. *Mem. Inst. Osw. Cruz.* 1992; 87: 241.

Yu, DB, Li, Y, Sleigh, AC, Yu, XL, Li, YS, Wei, WY, Liang, YS and McManus, DP. Efficacy of praziquantel against *Schistosoma japonicum*: field evaluation in an area with repeated chemotherapy compared with a newly-identified endemic focus in Hunan, China. *Trans.Roy.Soc. Trop.Med.Hyg.* 2001; 95: 537-41.

Zhang R, Takahashi S, Orita S, Yoshida A, Maruyama H, Shirai T, Ohta N. p53 gene mutations in rectal cancer associated with schistosomiasis japonica in Chinese patients. *Cancer Lett.* 1998;131:215-21.

Zheng J, Gu XG, Xu YL, Ge JH, Yang XX, He CH, Tang C, Cai KP, Jiang QW, Liang YS, Wang TP, Xu XJ, Zhong, JH, Yuan HC, Zhou XN. Relationship between the transmission of schistosomiasis japonica and the construction of the Three Gorges Reservoir. *Acta Trop.* 2002; 82:147-56.

Chapter 12

HOST GENETICS AND SCHISTOSOMIASIS

Jeffrey M. Bethony[1] and Jeff T. Williams[2]
Department of Microbiology and Tropical Medicine, The George Washington University Medical Center, Washington, DC [2]Department of Genetics, Southwest Foundation for Biomedical Research, San Antonio, TX, USA

Key words: host genetics, segregation analysis, quantitative genetic analysis, complex genetics

There is considerable variation in schistosome infection and disease. Wide variation has been observed in the prevalence and intensity of infection between nearby endemic areas (Pugh and Gilles, 1978; Polderman, 1979; Scott et al., 1982; Polderman et al., 1985; Fulford et al., 1992). Severe disease occurs in a relatively small proportion of the infected population; less than 10% of infected individuals suffer severe hepatosplenic schistosomiasis mansoni. The importance of host genetics on the variation in infection and disease development in schistosomiasis has been difficult to assess because of the multiplicity of environmental factors that are also involved, e.g., vector density, vector distribution, parasite virulence, etc. However, as a result of increased availability of sequence data and the development of novel statistical methods, the last ten years has seen unprecedented research on the role of host genetics during schistosome infection. This chapter outlines many of these advances by phenotype studied (infection intensity, immune response, disease development) and the different study designs and statistical methods used to study these phenotypes.

1.   GENETICS OF INFECTION INTENSITY

The intensity of infection is the most studied phenotype for two reasons: (1) heavy levels of fecal egg excretion are thought to be associated with disease development, and (2) variation in egg counts is common and large. The distribution of fecal egg counts between hosts is highly overdispersed, with only a small proportion of hosts being heavily infected; 10% of the host population excretes 70% of the eggs (Anderson and May, 1985). For example, a study of *S. haematobium* in Ethiopia found that 50% of the eggs were excreted by 5% of the community (Bradley and May, 1978) while a Brazilian study of *S. mansoni* infection found that 6% of the community was responsible for the majority of egg output (de Lima e Costa et al., 1985). Another often-studied phenotype related to egg counts is reinfection. Reinfection studies generate several kinds of data: (i) incidence of reinfection, (ii) time to reinfection, (iii) and intensity of reinfection. These phenotypes are thought to indicate "susceptibility" or "resistance" to reinfection.

The starting point for the study of host genetics of infection intensity was the observation of differences in the levels of egg counts between ethnic groups (see Quinnell, 2003 for review). Differences in susceptibility to soil-transmitted helminth infections between ethnic groups have been noted since the early 20th century, when studies in the southern USA consistently showed a much higher prevalence and intensity of hookworm infection in people of European ancestry compared to those of African ancestry (Smillie and Keller, 1925; Keller et al., 1937). Similar investigations of schistosomiasis in Brazil suggested that African and Caucasian Brazilians had similar levels of prevalence and intensity of infection (Nunesmaia et al., 1975; Bina et al., 1978), but Caucasian Brazilians developed the "hepatosplenic form of infection" more frequently than Brazilians of African descent.

Another important observation suggesting an effect of genetics on susceptibility to infection is the aggregation of infection intensity within households; the members of some households have overall heavier levels of infection than members of neighbor households. The household aggregation of infection intensity has been reported for the soil-transmitted helminth infections: hookworm (Behnke et al., 2000), *Strongyloides stercoralis* (Conway et al., 1995), *Ascaris,* and *Trichuris* (Williams et al., 1974; Forrester et al., 1988; Forrester et al., 1990; Anderson et al., 1993; Chan et al., 1994). Much research also points to the importance of the household with respect to water-borne helminth infections such as schistosomiasis. This is because the clustering of domestic activities associated with water collection, storage, and usage (Watts et al., 1998) often results in the sharing of infective sites and water contact behaviors, exposing household members to similar risks of infection. In a study on the causes of the heavy infections in an endemic area of Brazil, Dessein and colleagues (1992) observed that heavy infections were clustered in certain families. Bethony et al. (2001) determined that sharing the same residence accounted for 28% of the variance in fecal egg counts for *S. mansoni* in an endemic area of Brazil. When other well-documented risk factors such as gender, water contact behavior, occupation, and income were included in the model, sharing the same residence still accounted for 23% of the variation in fecal egg counts. Taken together, these results implicate the household as an important composite measure of the complex interrelationships between shared genetic, demographic, socioeconomic, environmental, and behavioral factors that may influence the transmission of schistosomiasis.

After documenting household aggregation of a trait, the next step is to separate the components of shared residence into the effects shared genes (kinship) and shared environment (household). Quantitative genetics is often utilized at this point because its methods can distinguish the degree to which a phenotype is determined by shared genes, shared residence, and other well-documented covariates (e.g., age, sex, income).

There are several terms to describe the effect of shared genes studied in quantitative genetics: "additive", "polygenic", "multifactorial", or "complex". These terms refer to the analyses of phenotypes (traits) that are influenced by multiple genes, usually with a significant environmental component. The traits in quantitative genetics are usually on a continuous scale; i.e., the values are on a continuum (e.g., height, weight, etc.) as opposed to a discrete scale or interval. The evidence for the action of genes on a trait derives from the covariances (or correlations) of the trait between different classes of relatives (e.g., parent-offspring; sib-sib, grandparent-grandchild, etc.). The *a priori* expectation for the correlation between relative pairs is given by the coefficient of relatedness for the individuals, which represents a genome-averaged expectation (or probability) that two individuals will share an allele identical by descent. This probability also determines the extent to which additive genetic factors contribute to the phenotype. Heritability is usually the first quantitative value estimated by a quantitative genetic analysis and is defined as the proportion of variation directly attributable to genetic differences among individuals relative to the total variation in a population (Falconer and Mackay, 1996). A high heritability offers strongly suggestive evidence for the role of additive genetics on a trait.

The estimation of heritability requires the assembly of nuclear families, but is most powerful when used with large, multi-household extended families ("pedigrees") (Williams and Blangero, 1999). For example, using extended, multi-household pedigrees in a *S. mansoni* endemic area of Brazil, Bethony and coworkers (2002) found that host additive genetic factors consistently accounted for a significant proportion of the variation in schistosome fecal egg excretion rates: 43% in an unadjusted model and 40% in a model adjusted for age, gender, household monthly income, and water contact frequency. This was accomplished by using a maximum likelihood variance components analysis, which partitioned the total variance in fecal egg counts into (i) host additive genetics, (ii) shared environment (household), and (iii) host-specific environmental factors. Variance components analysis does not assume any specific distribution of gene effects (as done below for segregation analysis). Instead, it assesses the aggregate effect of host additive genetic factors on the variation in fecal egg counts. The notion that additive genetic effects influence the distribution of egg counts fits best with the notion of an immune-facilitated excretion of schistosome eggs from infected patients; i.e., schistosome eggs exit the human host through the requisite facilitation of functional immune responses and that the complex nature of this host-parasite interaction is best captured by a model of polygenic inheritance in which the aggregate effect of multiple loci are modeled on the trait.

Although genetic control of egg counts is likely to be polygenic, there may be detectable single (major) gene effects, and an alternative approach to quantitative genetic analysis is to use segregation analysis to examine evidence for major gene effects and determine their mode of inheritance. In a study on the causes of the heavy *S. mansoni* infections in an endemic area of Brazil, Dessein et al. (1992; 1999) observed that certain subjects appeared to be become reinfected at heavy levels even after several rounds of treatment. He then used segregation analysis to study this predisposition to reinfection. Segregation analysis is a statistical method that evaluates whether mathematical models that incorporate a major gene effect are better than sporadic or multi-gene models to explain phenotype distribution. The segregation analysis in this study was done by means of a model-based analysis strategy combining 2 successive steps. In step one, segregation analysis provided evidence for a co-dominant major gene, denoted as SM1. Under this major gene model, 3% of the population was homozygous negative and predisposed to very heavy infection levels, 68% were homozygous for the SM1 and resistant to reinfection, and 29% were heterozygous with an intermediate level of resistance. In the second step, a genome screen was carried out by means of a classical (model based) lod-score (the decimal logarithm of the likelihood ratio for linkage versus non-linkage) analysis that was based on the SM1 model, with SM1 mapped to the q31-q33 region of chromosome 5 (Marquet et al., 1996; Marquet et al., 1999). 5q31-q33 contains a number of genes that encode cytokines that may play an important role in regulation of the immune response against helminth parasites, in particular genes coding for interleukin-4 (IL-4), IL-5, IL-12, IL-13, and CSF-1R, and the interferon regulatory factor-1 (IRF-1). The existence of a locus of susceptibility to *S. mansoni* in 5q31-q33 was reproduced in a Senegalese population (Muller-Myhsok et al., 1997). It has also been reported that blood parasitemia in *P. falciparum* infections are controlled by a locus in the same 5q31-q33 region (Rihet et al., 1998). This region has been linked to loci related to IgE and eosinophilia production, i.e., a locus regulating IgE levels, and a locus controlling bronchial hyper-responsiveness in asthma. These results indicate that this region is likely to play an important role in susceptibility to infectious diseases and provided important insights into the mechanisms of pathogenesis.

An autosome-wide scan was performed on these same Brazilian families (Zinn-Justin et al., 2001) which utilized an extension of the weighted pairwise correlation to consider the role of two unlinked loci influencing the phenotype. The genome scan was used to search for a second locus, taking into account linkage to the first locus SM1 at 5q31-q33. Using this method, they confirmed the presence and the importance of SM1 influencing *S. mansoni* infection levels but then went on to identify several candidate regions that may play a "much less significant role" independently or not of SM1. Two additional regions of special interest were region q21-q23 on chromosome 1 (results independent of 5q3l-q33) and region p2l-q2l on chromosome 6 (results in interaction with 5q3l-q33). The chromosome 1 region between DlS248 and D1S484 contains the colony stimulating factor ligand (CSFL) gene, the receptor of which is located in 5q31-q33. The 6p21-q21 region is located between 2 regions of major interest. One is the major histocompatibility complex (MHC) region, and the other is the 6q22-q23 region, which contains the gene encoding for the interferon gamma receptor ligand binding chain (IFNGRl) and which was recently linked to a locus controlling the development of severe hepatic fibrosis due to *S. mansoni* infection, denoted as SM2 (Dessein et al., 1999).

2.   GENETICS OF THE HOST IMMUNE RESPONSE TO SCHISTOSOME INFECTION

The regulation of the immune response to schistosome infection has been a topic of great interest. Susceptibility to infection or reinfection does not appear to be due to a general failure of the developing humoral immune response because individuals infected with helminths have elevated levels of antibody against crude antigen extracts and recombinant schistosome proteins. A consistent observation of these studies is that individuals vary greatly in their cellular and humoral immune responses against different crude antigen preparations. Heterogeneity in immune responses between individuals may give us some clue as to the source of the marked variation in intensity of infection observed in individuals from the same endemic area.

Studies on Brazilians and Kenyans infected with *S. mansoni* have suggested that the ability of the host to recognize different epitopes and mount specific immune responses is genetically restricted. For example, Bethony et al. (1999) demonstrated that there were significant differences in IgG4 and IgE levels raised against adult-stage antigens Smp20.8 and Smp50 and the egg-stage antigen Smp40 between siblings from different Brazilian families. In Kenya, anti-schistosome IgE responses were restricted to only a few people whose immune system recognized a limited repertoire of antigens (Dunne et al., 1992). Although no genetic basis has been established for the Kenyan results, they are consistent with genetic restriction of host immunocompetence and antigen recognition.

Bethony et al. (submitted) further describe the genetic factors on the variation in isotype responses to crude and defined *S. mansoni* antigens in the multi-household extended pedigree from Brazil. The finding of large and significant heritabilities for levels of antibody isotype responses to crude and defined schistosome antigens indicates that additive genetics play an important role in the humoral immune response to infection. Because of the physical and functional relationships among these immune-related traits, Bethony et al. investigated if these strong additive genetic effects derived from common genes (i.e., pleiotropy), with this situation being manifest as a pattern of significant genetic correlations among these traits. The significant and positive genetic correlations observed between pairs of isotypes against crude and defined antigens from the same life cycle stage (e.g. the IgG1 response to Soluble Worm Antigen Protein (SWAP) and the recombinant adult antigen Smp 20.8) indicate that a common gene or suite of genes contributes to the correlated regulation of the antibody isotype response to crude and defined antigens from the same developmental stage (adult) of the parasite within the host. Furthermore, this pleiotropy is biologically important since between one half (SWAP IgG1) and as such much as three quarters (Smp20.8 IgG1) of the total phenotypic variance of these immunological traits are attributable to the additive effects. Conversely, the absence of genetic correlations between the isotypes responses to crude antigen extracts (SWAP and SEA) from different life cycle stages indicates that the humoral immune response to different developmental stages of the parasite within the host may arise from different sets of genes. Even more striking was that fecal egg counts and the isotype response to recombinant components of the egg such as Smp40 (a major egg protein) showed no pleiotropic effects, indicating egg counts are not controlled by the same genes that control the immune response to egg antigens. In short, the genetics of the immune response to schistosome infection may be as complex, multi-factorial, and compartmentalized as the infection itself. This is in keeping with much of the literature on helminth infection: (1) the succession of developmental stages of the helminth within the host, which bear stage specific antigens; and (2) the succession of distinct immunological compartments occupied by each developmental stage of the helminth within the host, including the idea that helminth immune evasion mechanisms are likely to be finely tuned to each of these micro environments. These results are important for the direction of future genetic studies of helminth infection as they indicate that the genetics of egg counts (the most common phenotype studied) may not reliably indicate the complexity of the genetics that influence the immune response to helminth infection.

As shown above, Dessein et al. (1999) observed that resistance to reinfection may depend on a major locus SM1, which was mapped to chromosome 5 in the q31-q33 region. The 5q31-q33 region comprises a number of cytokine genes or genes encoding cytokine receptors, which are critical for helper T-helper cell (Th) differentiation. This and the various reports showing that Th1 or Th2 lymphocytes play a central role in immunity to infectious pathogens led Rodrigues et al. (1996) to do a clonal analysis in resistant and susceptible people to test whether SM1 affected the differentiation of Th1 or Th2 subsets. The fact that cytokine ratios follow a rather continuous distribution between resistant and susceptible subjects was used to determine if the presence of several alleles of susceptibility produced more than one susceptible genotype. Genotypes were defined according to the biallelic model genes model predicted by segregation analysis described above (Dessein et al., 1999); hence, subjects were either homozygous for the allele that determines resistance to infection or were homozygous for the allele that determines high rates of infection. Production of IL-4, IL-5, and IFN-γ by clones from these two groups was then assessed. T cell clones (TCC) specific for schistosome extracts from the two study groups (homozygous or heterozygous) differed markedly in two aspects: (1) TCC from homozygous resistant subjects produced significantly more IL-4 and more IL-5 than TCC from homozygous susceptible individuals; and (2) the pattern of IL-4, IL-5, and IFN-γ secretion was characteristic of each group, that is, resistant subjects yielded clones producing more IL-4 and IL-5 than IFN-γ, whereas IFN-γ was produced in greater quantities than IL-4 and IL-5 by TCC from susceptible subjects. These findings are consistent with the conclusion that TCC specific for schistosome antigens of resistant and susceptible individuals differ in IL-4 and IL-5 production, but not in IFN-γ production. The observation that TCC from resistant individuals are Th2 when they are specific for schistosome antigen, but Th1 when they are not specific for parasite antigens confirms that schistosome antigens may have the selective ability to produce the appropriate cytokine environment for Th2 development. This effect may be so marked that a normal Th1 response to a conventional antigen may, in certain subjects, deviate toward a Th2 response. Because Rodrigues and colleagues (1996) linked this property to the 5q31-q33 region, they hypothesized that the differences in the cytokine environment in resistant and susceptible subjects could be produced by certain polymorphisms in IL-4 or IL-13 genes that could modify the level of production of these cytokines. Together, these findings suggest that SM1 could play a key role in other infectious diseases that are dependent on the balance between type 1 and type 2 cytokines.

3. GENETICS OF PATHOLOGY

The phenotype in these studies is the severe clinical manifestations such as liver fibrosis, spleen congestion, and portal hypertension resulting from chronic periportal inflammation that occurs in between 3 and 10% of the infected population. This fibrosis is actually a part of the repair process that follows tissue damage caused by inflammation, which is (in turn) the result of $CD4^+T$ cell-dependent granuloma formation to parasite eggs lodged in host tissues, particularly the liver. The reasons for development of severe clinical forms of schistosomiasis vary. It has long been hypothesized that severe schistosomiasis develops in subjects with the heaviest infections, but no general consensus exists. The reason that intensity of infection is not thought to simply correlate with pathology is due to several reasons: (1) severe fibrosis may causes a reduction in egg excretion, leading to an underestimate of parasite load in such patients, that is, patients with severe clinical forms excrete fewer EPG feces; (2) the disease takes several years to develop and may occur years after the peak infection intensity; and (3) different studies analyzed different clinical phenotypes (hepatosplenomegaly or fibrosis evaluated by ultrasound).

The vast literature on the role of host genetics to severe clinical forms of schistosomiasis starts with the well known study performed in Brazil by Bina et al. (1978) who reported that subjects from Bahia, in Northeastern Brazil, with African characteristics did not develop severe disease in spite of heavy infections. After this follows a number of case control studies that address the role of HLA genes in determining variable susceptibility to disease, with many well-defined and positive HLA associations (Salam et al., 1979; Ohta et al., 1987; Wishahi et al., 1989; Hafez et al., 1991; Hirayama et al., 1998; May et al., 1998; Waine et al., 1998; McManus et al. 2001). Of these, the most widely cited study is that done by Secor and colleagues (1996) in Bahia, who tested the correlation between severe hepatosplenic disease caused by S. *mansoni* and MHC class II alleles. The idea was that $CD4^+$ T cells are centrally involved; either by producing the cytokines that regulate resistance to infection or by generating increased responses to egg antigens in patients who are in the process of developing hepatosplenic disease. Because $CD4^+$ T cell responses are dependent on antigen presentation in the context of the class II major histocompatibility complex (MHC), a patient's MHC class II genotype may exert an influence on susceptibility to infection, regulation of immune responsiveness to egg antigens, and/or progression to hepatosplenic disease. A strong association was found between HLA-DQB1-0201 and hepatosplenic disease. Furthermore, patients positive for DRB1*01, DQA1*0101, or DQB1*0501 were less likely to respond to SEA than the study population. However, none of the associations with SEA responsiveness remained significant after Bonferoni correction.

Ultrasound evaluations of 800 subjects living in an endemic area of Sudan showed that severe fibrosis associated with portal hypertension was frequent in certain families and absent in others despite the fact that all families had been living for years in the same exposure conditions (Mohamed-Ali et al., 1999). Subjects with stage 2 (FII) and stage 3 (FIII) fibrosis and evidence of portal hypertension were classified as affected and segregation analysis was applied to this phenotype. The analysis showed evidence for a major locus in 6q22-q23 (Dessein et al., 1999), very close to the gene encoding the alpha 2 chain of the IFN-γ receptor, a result consistent with the well known anti-fibrogenic properties of IFN-γ. Furthermore, a number of studies have demonstrated that polymorphisms in the genes encoding either one of the two chains of the receptor for IFN-γ increases human susceptibility to infectious diseases in which the host immune response is characterized by a persisting granulomatous reaction such as schistosomiasis and tuberculosis (Pierre-Audigier et al., 1997; Roesler et al., 1999). To determine whether gene SM2 controlling disease development and SM1 that controls infection levels were distinct, Dessein and colleagues went on to test whether SM2 was linked to the 5q31-q33 genetic region. These studies showed that two distinct genetic loci control human susceptibility to *S. mansoni*: SM1 located in 5q31-q33 controls infection levels, probably by acting on the production of certain cytokines, while SM2 controls disease progression. This result does not rule out an interaction between SM1 and SM2, but it is reasonable to postulate that disease development is accelerated in SM2-predisposed subjects by heavy infections.

4.   DISCUSSION

Analysis of the genetic basis of diseases like schistosomiasis is indeed "complex" for a number of reasons. Not only is this a highly polygenic disease with important, if not overwhelming, genetic components, but there is the role of the parasite as well. Nonetheless, steady progress has been made over the last ten years to untangle the complex interplay of host and parasite genes that results in some striking inter-individual variation in infection intensity and disease. However, greater consensus is needed before the entire puzzle can be "pieced together."

The first area were consensus is needed is the phenotype variously referred to as "fecal egg counts, "infection intensity", or "fecal egg excretion rates". Egg counts are obtained by examination of fecal samples for the presence of ova, typically using the Kato Katz thick smear technique. This is the method of choice as it is relatively straightforward, allows a large number of samples to be examined within a short time, and provides a quantitative measure of infection intensity. However, variation introduced by the technique itself needs to be considered. The Kato Katz thick smear technique has been the subject of much scrutiny over the years; with much of this scrutiny bearing on the day-to-day variation in egg counts in the same individual. Furthermore, egg counts in the Kato Katz method vary as a function of the location of the sampling within a single stool specimen.

Studies in Brazil (Barreto et al., 1990) showed that intra-specimen variation explained 31.5 % of the total egg count variance, with day to day variation accounting for 46.1% of the variation. The argument given by geneticists is that, given all of this variation in the phenotype, the detection of a strong and significant pattern of genetic influence is even more striking, as the probably of such phenotype having a genetic pattern in the face of variation introduced by the method is highly unlikely. However, the problem remains of how to interpret the strong genetic patterns observed in egg counts from individuals living in endemic areas. It has yet to be shown that egg counts derived from the Kato Katz thick smear technique are related in a linear manner to adult worm burden. It is (at best) an approximate measure of how many worms are harbored by infected humans, since the direct quantification of *S. mansoni* adult worms in endemic situations is impossible.

The second area were consensus is needed is the assumption that a single gene or a single suite of genes controls the entire infectious process. In both animals and human immune studies, the complexity of the schistosome life cycle within the host offers numerous opportunities for the parasite and host to interact not only at different molecular levels, but also within diverse immune microenvironments: e.g., during skin invasion, penetration of the gut mucosa, and adults dwelling in the gut. Hence, not only is there diversity in developmental stage of the parasite within the host, but also regional specialization of host immune response (skin, lung, mucosal), which results in a diverse set of antigenic challenges, occurring in different immunological niches. In short, it may be that the genetics of egg counts may simply reflect the complex passage of the egg into the feces or some other aspect of fecundity. One only need to look at the poor relationship between egg counts and heptosplenic disease to note that egg counts are not an informative phenotype of the severity of schistosomiasis pathology. The observation that the major locus controlling fibrosis is not linked to chromosome 5q31-33 demonstrates that anti-disease immunity and anti-infection immunity are under distinct major gene controls.

A third aspect to be considered is that some of the variations in immune responses might reflect the different methodologies used by various research groups. Standardization of the methodologies used to derive immune phenotypes is needed, as suggested by Mutapi (2001). An area of special concern is the antigen preparations used in these assays. Ideally, single, stage specific recombinant antigens should be used. While crude antigens are valuable for detecting general patterns in infected populations, single antigens may permit a better definition of humoral and cellular responses under investigation. The difference may be related to the fact that crude antigen extracts of the parasite contain large populations of proteins, which present numerous and diverse epitopes to the immune system. In comparison, defined antigens that consist of a single polypeptide present the immune system with a more limited set of epitopes. As such, one might anticipate that immune responsiveness to a crude *S. mansoni* antigen preparation would be more complex and involve interactions of antigen peptides with several class II molecules.   Conversely, responsiveness to defined antigens would involve fewer major T cell epitopes, with fewer class II molecules.   Studies highly polymorphic human populations suggest that more complex protein mixtures may not yield informative data (Marsh and Huang, 1991; Marsh et al., 1991).   Thus, in the analysis of the genetic influence on immune responsiveness, defined antigens may prove more informative than crude preparations.   Furthermore, reported differences in the immunogenicity of antigens from different parasite strains, suggest the immune responses of human populations to antigens generated from allopatric parasite strains should be treated with caution (Mutapi 2001).

Finally, studies on the ability of the host to recognize antigens and to mount an immune response have generally been performed before treatment. There are currently no studies specifically on the effect of host genetics on post-treatment antibody changes (Mutapi 2001).   While drug treatment does not modify the host genetic make-up, treatment of infected people alters schistosome-specific immune responses, and these changes are thought to confer some resistance to re-infection and protect against the development of severe disease forms.   Host genetics influence levels of antibody isotype to different crude and defined antigens from *S. mansoni* as well as affect the differentiation of parasite-specific Th2 cells (Rodrigues et al., 1996; Bethony et al., 1999). Hence, it is important to understand the role that host genetics may play in influencing the immune responses after praziquantel treatment in different populations.   This research will have important implications for vaccine development and evaluation for two reasons: (1) the majority of vaccine target populations will have already been treated as part of national control or research programs, and (2) most vaccine candidates to date have recombinant proteins antigens (e.g. 28GST, the Bilvax vaccine).

In the near future, two new approaches will become of increasing importance for the study of the host genetics of schistosomiasis.   The first will be the rise of detailed linkage disequilibrium mapping, resulting from the availability of huge numbers of new SNPs in all areas of the genome. This will initially allow much more precise mapping of known associations and linkages and eventually lead to genome-wide association studies.   The potential of the latter has been extensively discussed in the field of complex disease genetics in general.   The apparently highly polygenic nature of common infectious disease suggests that this approach may be particularly fruitful in this arena.   An approach that is especially exciting is combining host-parasite genetic analysis.   In diseases where it is readily possible to sample the genomes of both host and pathogen, such as HIV, malaria, and many other infections, this is beginning to lead to new combined analytical approaches that may reveal the nature of evolutionary driving forces for host and parasite genetic diversity. However, investigating interactions may be more feasible for organisms like toxoplasmosis, malaria, or viruses as infections are frequently clonal and the interaction is between one host genome and one parasite genome can be studied. However, helminth infections contain numerous genotypes in a single infected host, since there is no direct reproduction in the host; as such, the analysis must involve a single host and multiple worm genotypes. A more feasible approach to studying host-helminth interactions is the use of inbred lines of parasites.

REFERENCES

Anderson, R. M. and R. M. May (1985). "Helminth infections of humans: mathematical models, population dynamics, and control." Adv Parasitol 24: 1-101.

Anderson, T. J., C. A. Zizza, et al. (1993). "The distribution of intestinal helminth infections in a rural village in Guatemala." Mem Inst Oswaldo Cruz 88(1): 53-65.

Barreto, M. L., D. H. Smith, et al. (1990). "Implications of faecal egg count variation when using the Kato-Katz method to assess *Schistosoma mansoni* infections." Trans R Soc Trop Med Hyg 84(4): 554-5.

Behnke, J. M., D. De Clercq, et al. (2000). "The epidemiology of human hookworm infections in the southern region of Mali." Trop Med Int Health 5(5): 343-54.

Bethony, J., A. M. Silveira, et al. (1999). "Familial resemblance in humoral immune response to defined and crude *Schistosoma mansoni* antigens in an endemic area in Brazil." J Infect Dis 180(5): 1665-73.

Bethony, J., J. T. Williams, et al. (2002). "Additive host genetic factors influence fecal egg excretion rates during *Schistosoma mansoni* infection in a rural area in Brazil." Am J Trop Med Hyg 67(4): 336-43.

Bethony, J., J. T. Williams, et al. (2001). "Exposure to *Schistosoma mansoni* infection in a rural area in Brazil. Ii: household risk factors." Trop Med Int Health 6(2): 136-45.

Bina, J. C., J. Tavares-Neto, Et Al. (1978). "Greater resistance to development of severe schistosomiasis in Brazilian Negroes." Hum Biol 50(1): 41-9.

Bradley, D. J. And R. M. May (1978). "Consequences of helminth aggregation for the dynamics of schistosomiasis." Trans R Soc Trop Med Hyg 72(3): 262-73.

Chan, L., D. A. Bundy, et al. (1994). "Genetic relatedness as a determinant of predisposition to *Ascaris lumbricoides* and *Trichuris trichiura* infection." Parasitology 108 (pt 1): 77-80.

Conway, D. J., A. Hall, et al. (1995). "Household aggregation of *Strongyloides stercoralis* infection in Bangladesh." Trans R Soc Trop Med Hyg 89(3): 258-61.

De Lima E Costa, M. F., R. S. Rocha, et al. (1985). "A clinico-epidemiological survey of *Schistosomiasis mansoni* in a hyperendemic area in Minas Gerais state (Comercinho, Brazil) I: Differences in the manifestations of schistosomiasis in the town centre and in the environs." Trans R Soc Trop Med Hyg 79(4): 539-45.

Dessein, A. J., P. Couissinier, et al. (1992). "Environmental, genetic and immunological factors in human resistance to *Schistosoma mansoni*." Immunol Invest 21(5): 423-53.

Dessein, A. J., D. Hillaire, et al. (1999). "Severe hepatic fibrosis in *Schistosoma mansoni* infection is controlled by a major locus that is closely linked to the Interferon-gamma receptor gene." Am J Hum Genet 65(3): 709-21.

Dessein, A. J., S. Marquet, et al. (1999). "Infection and disease in human *Schistosomiasis mansoni* are under distinct major gene control." Microbes Infect 1(7): 561-7.

Dunne, D. W., A. E. Butterworth, et al. (1992). "Human IgE responses to *Schistosoma mansoni* and resistance to reinfection." Mem Inst Oswaldo Cruz 87 suppl 4: 99-103.

Forrester, J. E., M. E. Scott, et al. (1988). "Clustering of *Ascaris lumbricoides* and *Trichuris trichiura* infections within households." Trans R Soc Trop Med Hyg 82(2): 282-8.

Forrester, J. E., M. E. Scott, et al. (1990). "Predisposition of individuals and families in mexico to heavy infection with *Ascaris lumbricoides* and *Trichuris trichiura.*" Trans R Soc Trop Med Hyg 84(2): 272-6.

Fulford, A. J., A. E. Butterworth, et al. (1992). "On the use of age-intensity data to detect immunity to parasitic infections, with special reference to *Schistosoma mansoni* in Kenya." Parasitology 105 (pt 2): 219-27.

Hafez, M., S. Aboul Hassan, et al. (1991). "Immunogenetic susceptibility for post-schistosomal hepatic fibrosis." Am J Trop Med Hyg 44(4): 424-33.

Hirayama, K., H. Chen, et al. (1998). "Glycine-valine dimorphism at the 86th amino acid of HLA-DRB1 influenced the prognosis of postschistosomal hepatic fibrosis." J Infect Dis 177(6): 1682-6.

Keller, A., W. Leathers, et al. (1937). "The present status of hookworm infestation in North Carolina." Am J Hyg 26: 432-454.

Marquet, S., L. Abel, et al. (1999). "Full results of the genome-wide scan which localises a locus controlling the intensity of infection by *Schistosoma mansoni* on chromosome 5q31-q33." Eur J Hum Genet 7(1): 88-97.

Marquet, S., L. Abel, et al. (1996). "Genetic localization of a locus controlling the intensity of infection by *Schistosoma mansoni* on chromosome 5q31-q33." Nat Genet 14(2): 181-4.

Marsh, D. G. and S. K. Huang (1991). "Molecular genetics of human immune responsiveness to pollen allergens." Clin Exp Allergy 21 suppl 1: 168-72.

Marsh, D. G., P. Zwollo, et al. (1991). "Molecular and cellular studies of human immune responsiveness to the short ragweed allergen, Amb A V." Eur Respir J Suppl 13: 60s-67s.

May, J., P. G. Kremsner, et al. (1998). "HLA-DP control of human *Schistosoma haematobium* infection." Am J Trop Med Hyg 59(2): 302-6.

Mcmanus, D. P., A. G. Ross, et al. (2001). "HLA class II antigens positively and negatively associated with hepatosplenic schistosomiasis in a Chinese population." Int J Parasitol 31(7): 674-80.

Mohamed-Ali, Q., N. E. Elwali, et al. (1999). "Susceptibility to periportal (symmers) fibrosis in human *Schistosoma mansoni* infections: evidence that intensity and duration of infection, gender, and inherited factors are critical in disease progression." J Infect Dis 180(4): 1298-306.

Muller-Myhsok, B., F. F. Stelma, et al. (1997). "Further evidence suggesting the presence of a locus, on human chromosome 5q31-q33, influencing the intensity of infection with *Schistosoma mansoni.*" Am J Hum Genet 61(2): 452-4.

Mutapi, F. (2001). "Heterogeneities in anti-schistosome humoral responses following chemotherapy." Trends Parasitol 17(11): 518-24.

Nunesmaia, H. G., E. S. Azevedo, et al. (1975). "[Racial composition and ahaptoglobinemia in carries of the hepatosplenic form of *Schistosomiasis mansoni*]." Rev Inst Med Trop Sao Paulo 17(3): 160-3.

Ohta, N., M. Hayashi, et al. (1987). "Immunogenetic factors involved in the pathogenesis of distinct clinical manifestations of *Schistosomiasis japonica* in the Philippine population." Trans R Soc Trop Med Hyg 81(2): 292-6.

Pierre-Audigier, C., E. Jouanguy, et al. (1997). "Fatal disseminated *Mycobacterium smegmatis* infection in a child with inherited Interferon gamma receptor deficiency." Clin Infect Dis 24(5): 982-4.

Polderman, A. M. (1979). "Transmission dynamics of endemic schistosomiasis." Trop Geogr Med 31(4): 465-75.

Polderman, A. M., K. Mpamila, et al. (1985). "Historical, geological and ecological aspects of transmission of intestinal schistosomiasis in Maniema, Kivu Province, Zaire." Ann Soc Belg Med Trop 65(3): 251-61.

Pugh, R. N. and H. M. Gilles (1978). "Malumfashi endemic diseases research project, III. urinary schistosomiasis: a longitudinal study." Ann Trop Med Parasitol 72(5): 471-82.

Quinnell, R. J. (2003). "Genetics of susceptibility to human helminth infection." Int J Parasitol 33(11): 1219-31.

Rihet, P., Y. Traore, et al. (1998). "Malaria in humans: *Plasmodium falciparum* blood infection levels are linked to chromosome 5q31-q33." Am J Hum Genet 63(2): 498-505.

Rodrigues, V., Jr., L. Abel, et al. (1996). "Segregation analysis indicates a major gene in the control of Interleukine-5 production in humans infected with *Schistosoma mansoni*." Am J Hum Genet 59(2): 453-61.

Roesler, J., B. Kofink, et al. (1999). "*Listeria monocytogenes* and recurrent mycobacterial infections in a child with complete interferon-gamma-receptor (INF-Gamma-R-1) deficiency: mutational analysis and evaluation of therapeutic options." Exp Hematol 27(9): 1368-74.

Salam, E. A., S. Ishaac, et al. (1979). "Histocompatibilty-linked susceptibility for hepatospleenomegaly in human schistosomiasis mansoni." J Immunol 123(4): 1829-31.

Scott, D., K. Senker, et al. (1982). "Epidemiology of human *Schistosoma haematobium* infection around Volta Lake, Ghana, 1973-75." Bull World Health Organ 60(1): 89-100.

Secor, W. E., H. Del Corral, et al. (1996). "Association of hepatosplenic schistosomiasis with HLA-DQB1*0201." J Infect Dis 174(5): 1131-5.

Smillie, W. And D. Keller (1925). "Intensity of hookworm infection in Alabama: its relationship to residence, occupation, age, sex, and race." J Am Med Ass 85: 1958-1963.

Waine, G. J., A. G. Ross, et al. (1998). "HLA class II antigens are associated with resistance or susceptibility to hepatosplenic disease in a Chinese population infected with *Schistosoma japonicum*." Int J Parasitol 28(4): 537-42.

Watts, S., K. Khallaayoune, et al. (1998). "The study of human behavior and schistosomiasis transmission in an irrigated area in Morocco." Soc Sci Med 46(6): 755-65.

Williams, D., G. Burke, et al. (1974). "Ascariasis: a family disease." J Pediatr 84(6): 853-4.

Williams, J. T. And J. Blangero (1999). "Power of variance component linkage analysis to detect quantitative trait loci." Ann Hum Genet 63(pt 6): 545-63.

Wishahi, M., H. G. El-Baz, et al. (1989). "Association between HLA-A, B, C and Dr antigens and clinical manifestations of *Schistosoma haematobium* in the bladder." Eur Urol 16(2): 138-43.

Zinn-Justin, A., S. Marquet, et al. (2001). "Genome search for additional human loci controlling infection levels by *Schistosoma mansoni*." Am J Trop Med Hyg 65(6): 754-8.

Chapter 13

CURRENT AND FUTURE ANTI-SCHISTOSOMAL DRUGS

Donato Cioli and Livia Pica-Mattoccia
Institute of Cell Biology, National Research Council, 00016 Monterotondo (Roma), Italy

Key words: antischistososmal drugs; oxamniquine; praziquantel

1.   INTRODUCTION

The chemotherapy of schistosomiasis poses intellectual challenges that have little to do with the boring series of blind trial-and-errors of the past, but have rather to be addressed with the most advanced tools of cutting-edge biology. And, not a trivial detail, chemotherapy is still the only approach that offers some immediate and tangible benefit to the many millions of people that are infected with schistosomes.

We shall try to support the above statements looking at the two drugs (oxamniquine and praziquantel) that are currently the only ones to be commercially available against schistosomiasis. The remaining antischistosomal drugs have been analyzed in previous reviews (Cioli *et al.*, 1995; Cioli, 1998).

2.   OXAMNIQUINE

Introduced in clinical practice in the early 1970s, oxamniquine has gained itself a well deserved reputation of safety and effectiveness. It is remarkable that such a success has been accorded to a compound that is the direct chemical descendant of a previous antischistosomal drug, hycanthone, that was dismissed and abandoned with the ignominious accusation of being a potent mutagen, a potential carcinogen and a suspected teratogen. A look at Fig 1, will convince of the obvious structural similarities between the two drugs (thick lines) and will also show that hycanthone has a 3-ring planar structure typical of DNA-intercalating agents (like acridine dyes), whereas oxamniquine has a simpler structure and has been proven to be devoid of intercalating activity (Pica-Mattoccia *et al.*, 1989). Thus, while hycanthone has become a model for intercalating drugs that produce frameshift mutations (Waring, 1970), oxamniquine has passed all laboratory tests of mutagenicity and –most important– all test of clinical safety during three decades of administration to millions of patients.

Hycanthone   Oxamniquine

The major limitation of oxamniquine is that its activity is practically confined to *S. mansoni* infections, while *S. haematobium* and *S. japonicum* are not sensitive to the drug. As a consequence, oxamniquine has been widely used in South America, where only *S. mansoni* is present; while in different parts of the world where infections with the other species are present, a wider preference has been accorded to praziquantel that is effective against all schistosomes. This geographical limitation did not encourage market competition among oxamniquine producers, so that its price has remained practically unchanged, while the opposite occurred for praziquantel, whose price underwent a dramatic reduction. Thus, even in countries like Brazil, where oxamniquine has traditionally been the antischistosomal drug of choice, price considerations are leading to a progressive decrease of its use.

2.1   Some features of oxamniquine activity

The effects of oxamniquine on schistosomes are delayed in time. When worms are exposed to the drug *in vitro*, no immediate adverse effects are noted; in fact, there is a remarkable stimulation of their motor activities, due to some anti-cholinergic effects that are not connected with the antiparasitic action (Pica-Mattoccia & Cioli, 1986). It is only 5-7 days after a 30-min *in vitro* exposure and subsequent culture in normal medium that schistosomes begin to show some damage, which eventually progresses to complete death in a few more days.

The other remarkable feature of oxamniquine (and of hycanthone) is that the length of schistosome exposure to the compound can be very short: about 15 minutes of contact *in vitro* or about 2 hours *in vivo* are perfectly adequate to cause a complete (and delayed) death of parasites after transfer to drug-free medium or to untreated animals (Cioli & Knopf, 1980).

Oxamniquine is more effective against male worms than it is against females, both *in vivo* and *in vitro* (by roughly a factor of 2). Also, oxamniquine has very little activity against immature schistosomes (worms that are 1- to 4-week-old).

Structure-activity studies show that the hydroxymethyl ($-CH_2OH$) function of oxamniquine and hycanthone is absolutely essential for drug activity. Analogs with a simple methyl group ($-CH_3$) are inactive *in vitro* or upon parenteral injection, but can be active upon oral administration since the host intestinal flora converts them to the hydroxymethyl metabolites. Compounds with a simple hydrogen atom in this critical position have no activity whatsoever.

We shall try to offer an explanation for all these features.

2.2   Oxamniquine resistance

Resistance to oxamniquine was described very early in the laboratory (Rogers & Bueding, 1971) and was soon confirmed in the field (Katz *et al.*, 1973). Schistosomes resistant to oxamniquine are also completely resistant to hycanthone (side-resistance), as one would expect from the structural similarities.

The magnitude of resistance to hycanthone/oxamniquine is quite striking, since resistant schistosomes can survive drug doses that are almost 1000-fold higher than the dose which is effective against sensitive parasites (Pica-Mattoccia *et al.*, 1993).

The most surprising fact about oxamniquine resistance is that, in spite of the sharp insusceptibility conferred, it never became a real public health problem, but remained confined to a few sporadic cases, without any apparent tendency to spread in the population.

2.3   Resistance is a recessive character

The most relevant information leading to the clarification of the mechanism of action of oxamniquine was obtained from genetics. The availability of schistosomes displaying a strong resistance to the drug was, in this context, a fortunate circumstance. By performing genetic crosses between sensitive and resistant schistosomes, it was possible to establish that resistance to oxamniquine is a recessive character. A single sensitive male and a single resistant female (and vice-versa) were introduced with a simple surgical cannulation into the mesenteric veins of a mouse. Since schistosomes possess very sophisticated mechanisms for finding each other, the two worms would soon mate and produce hybrid eggs. Adult schistosomes derived from these eggs were exposed to oxamniquine *in vitro* to decide whether they were sensitive or resistant to the drug. The results were totally unequivocal: all the hybrid schistosomes of the first generation (F1) were completely sensitive to oxamniquine, irrespective of whether the resistant parent was the male or the female (Cioli *et al.*, 1992). Such a simple Mendelian autosomal recessive inheritance was confirmed in the F2 generation (where about 25% of the progeny was resistant) and in backcrosses with the resistant parent (where about 50% of the progeny was resistant).

What are the implications of the above findings? In a diploid organism like the schistosome, a recessive character means that both chromosomes must carry the trait in order for the phenotype to appear. If a single chromosome has the 'mutation' (as expected in the F1 hybrids), the other one can usually compensate and produce the critical factor in sufficient amounts to result in a normal phenotype. In other words, recessivity usually indicates the lack of a function. Hence the obvious suggestion that resistant schistosomes may be missing some function necessary for the expression of oxamniquine activity, e.g. some factor that 'activates' the drug.

2.4   Oxamniquine alkylates the DNA of sensitive schistosomes

A particularly compelling analogy seemed to exist between oxamniquine and some known pro-carcinogens (like dimethylbenzanthracene) that are transformed into the carcinogenic substance once they are sulfonated (i.e. transformed into sulfate esters) under the activity of host sulfotransferases (Watabe *et al.*, 1982). Sulfotransferases are enzymes that transfer a sulfate group from a donor molecule (phosphoadenosinephosphosulfate, PAPS) to the hydroxyl group of an acceptor substrate. Oxamniquine could indeed function as such an acceptor, since –as we have seen– it possesses a critical hydroxyl group.

In many cases, the organic esters formed by sulfation are unstable ones and tend to dissociate spontaneously, thus giving rise to positively charged moieties (electrophiles) that have a strong tendency to react with other molecules and to form with them covalent bonds (alkylation reaction). From the dissociation of the dimethylbenzoanthracene sulfate, a reactive ion is generated that tends to alkylate DNA and other macromolecules, eventually leading to mutation and cancer. It was tempting to imagine that from the dissociation of an oxamniquine sulfate an alkylating agent could be generated that could alkylate schistosome macromolecules eventually leading to parasite death. The ensuing paragraphs illustrate the evidence that was collected to support such a hypothesis.

Tritiated oxamniquine was produced and it was shown that, upon a short contact with schistosomes, significant amounts of drug would form stable covalent bonds with worm DNA and other macromolecules. This occurred in the case of sensitive schistosomes, whereas oxamniquine resistant worms failed to bind any significant amount of tritiated drug. Also interestingly, no oxamniquine-DNA binding occurred in the species that are intrinsically resistant to the drug, like *S. haematobium* and *S. japonicum* (Pica-Mattoccia *et al.*, 1989).

2.5　Artificial drug esters alkylate the DNA of sensitive and resistant schistosomes

If drug activity is absent in resistant worms because the activating enzyme is not present and the oxamniquine ester cannot be formed, one would predict that a pre-formed drug ester should be able to exert activity even in drug resistant schistosomes. It was not possible to test sulfate esters, since these proved to be too unstable, but it was possible to show that, at least with hycanthone, the acetate, the methylcarbamate and other more stable esters were indeed able to form covalent bonds with the DNA of resistant schistosomes and produce their death, both *in vitro* and *in vivo* (Archer *et al.*, 1988).

2.6　An enzymatic activity leading to drug-DNA binding is absent in resistant schistosomes

An extract of sensitive schistosomes was found to contain an enzymatic activity capable of producing *in vitro* the covalent binding of tritiated oxamniquine (or tritiated hycanthone) to DNA and other macromolecules. Such an activity was absent in resistant worms, in *S. japonicum*, in *S. haematobium* and, most important, in human and rat liver, as well as in HeLa cells. The latter finding clearly provides a clue to oxamniquine species specificity, i.e. to its harmful effects to schistosomes and not to humans. We also found that the drug-DNA binding activity was less abundant in female worms and scarcely present in immature schistosomes, in accordance with the drug sensitivity of these organisms (Pica-Mattoccia *et al.*, 1992, 1997).

2.7   The drug-DNA binding activity is due to a sulfotransferase

When using partially purified worm fractions, the above enzymatic reaction requires the addition of the universal sulfate donor, PAPS. In addition, an extract from sensitive worms was able to sulfonate β-estradiol and quercetin (two typical sulfotransferase substrates), whereas an extract from oxamniquine resistant schistosomes failed to sulfonate the above substrates (unpublished). When a schistosome extract was separated by gel filtration on a Sephadex column, the elution of drug-DNA binding activity corresponded to a MW of about 30,000, which is compatible with the size expected for the subunit of a typical sulfotransferase. From the above evidence, we conclude that the schistosome enzyme leading to oxamniquine activation is most likely a sulfotransferase (Cioli e*t al.*, 1995).

2.8   Proposed mechanism of oxamniquine activity

Oxamniquine is a pro-drug that, under the activity of a specific sulfotransferase present in sensitive schistosomes, is converted within the parasite to a sulfate ester, the point of attack being at the hydroxymethyl group that we have seen to be essential for activity. The ester undergoes spontaneous dissociation and gives rise to an alkylating moiety capable of forming covalent bonds with schistosome macromolecules, including DNA (Cioli *et al.*, 1985). This reaction is a rapid one, since the ester is likely to be very unstable and reactive: this explains the short time of contact that is needed for oxamniquine activity. Parasite death, however, does not occur immediately, but the alkylation of nucleic acids leads to a progressive blockade of schistosome cell division and protein synthesis, until death takes place a few days later. Since, as we have seen, the activating sulfotransferase is not present at the same level in the two sexes and at various developmental stages, the drug sensitivity of these organisms is correspondingly different. The fact that immature schistosomes have very low levels of activating enzyme could be explained with a delayed expression of the corresponding gene in the developmental program of the worm. Oxamniquine resistant schistosomes have lost the sulfotransferase activity (through mechanisms that might go from a single nucleotide substitution to more complex genetic events) and thus escape macromolecule alkylation and subsequent death. It seems, however, that the lack of a functional sulfotransferase has a "cost" in terms of fitness for the resistant parasites. This was directly demonstrated by measuring various parameters of vitality in the schistosome life cycle. Decreased fitness, plus the recessive mechanism of inheritance, may very well explain why oxamniquine resistant schistosomes did not show a tendency to spread in the human population.

2.9   A future for oxamniquine

This is an unfinished story. The gene for the schistosome sulfotransferase that produces oxamniquine activation has not been cloned, yet, but the rapid progress of sequencing programs gives good hope that the gene will be soon discovered. This may open quite interesting prospects for the future of oxamniquine. First of all, once the structure of the activating enzyme is known in detail for both sensitive and resistant schistosomes, it should be possible to design oxamniquine derivatives that would have a good affinity for the binding site of the 'resistant' enzyme as well, thus overcoming the resistance problem. More important, though, would be the definition of the structure of the sulfotransferases of *S. haematobium* and *S. japonicum*, so that oxamniquine analogs active in these species could be designed. This would overcome the major limitation of oxamniquine, i.e. its restriction to *S. mansoni* infections, and would allow the continuing exploitation of the fortunate circumstance represented by the existence of an enzymatic activity that appears to be uniquely present in the parasite.

3.   PRAZIQUANTEL

Currently the drug of choice for the treatment of all forms of schistosomiasis, praziquantel was introduced in clinical practice in the late 1970s. Praziquantel is also active against other trematodes (*Opistorchis, Paragonimus, Fasciolopsis, Heterophyes, Metagonimus*) and against cestodes of human and veterinary importance. Its structure is based on a pyrazino-isoquinoline ring system that was initially explored as a possible source of tranquillizers. It contains an asymmetric carbon atom (asterisk in Fig. 2), which gives rise to the existence of two enantiomers, usually designated as *dextro-* and *laevo-* praziquantel. The *laevo*-enantiomer possesses most of the antischistosomal activity, while the *dextro*-enantiomer is almost inactive. As with many chiral drugs, however, the commercial preparations are a 50:50 racemate, because the isolation of the active isomer is a rather expensive process. This is regrettable, since it has been shown that patients treated with *laevo*-praziquantel at half the dose of the racemate had the same cure rate but fewer side effects (Wu *et al.*, 1991).

Praziquantel

3.1   Safety

There is a remarkable consensus on the general safety of praziquantel. Experimental evidence from a number of bacterial and mammalian systems showed no genotoxic effects (Kramers *et al.*, 1991) and animal tests confirmed the very low toxicity of the drug, both in acute and long-term experiments (Frohberg, 1984). The cumulated experience from the huge number of patients that have been treated up to this date clearly shows that praziquantel is a very well tolerated drug. Side effects are certainly present, but they are seldom serious and are usually short-lived. As a consequence, praziquantel is commonly administered without close medical supervision, e.g. by school teachers, in mass chemotherapy programs.

Up until very recently, praziquantel was not administered to pregnant and lactating women, but a close examination of the issue led to the conclusion that omitting the treatment entailed more risks than actually using it. An *ad hoc* committee convened in 2002 by the World Health Organization suggested extending treatment to pregnant and lactating women (Allen *et al.*, 2002).

3.2   Cost and availability

As already mentioned, the wide spectrum of activity of praziquantel encouraged its adoption in most endemic areas. As a consequence, praziquantel production underwent a rapid expansion and market competition resulted in a dramatic price drop. Today, generic praziquantel from several different producers is available on the market. A recent chemical survey on 34 drug samples collected at the user level in various countries, showed that generic products usually meet the required quality standards, but it showed also that two samples were completely fake since they contained no praziquantel at all (Sulaiman *et al.*, 2001). This can perhaps be taken as a sad testimonial to the popularity of praziquantel. It has been calculated that, in Egypt alone, about 20 million treatments have been administered between 1997 and 1999.

In spite of the relatively low price (about US$ 0.40 or less per treatment), drug cost is still a major limiting factor in many endemic countries, so that only a fraction of the infected people receives adequate treatment. To alleviate this problem, a recent program called "Schistosomiasis Control Initiative" is about to be launched with massive funding from the Gates Foundation, and is expected to provide praziquantel treatment to many millions of people in a few selected African countries.

3.3   Immuno-potentiation

Experiments carried out in the infected mouse have shown that the efficacy of praziquantel is significantly decreased in animals that are immuno-compromised (Sabah *et al.*, 1985). Drug efficacy can be restored by the injection of immune serum (Brindley & Sher, 1987). Praziquantel treatment increases the exposure of schistosome antigens at the parasite surface and some of these antigens have been identified and characterized (Doenhoff *et al.*, 1987; Sauma & Strand, 1990). These data are interpreted as evidence that the drug causes an initial damage to the parasite surface and that, as a consequence, parasite antigens are exposed that function as targets of the host immune response, thus potentiating praziquantel activity. No clear evidence exists regarding praziquantel immuno-potentiation in humans. A study on immunocompromised HIV-infected patients showed a normal level of praziquantel efficacy, but previous sensitization to schistosome antigens could not be excluded in these patients (Karanja *et al.*, 1998).

3.4   Parasite age

Since the very first publications on praziquantel activity, it was pointed out that the drug has very little effect on immature parasites (Gönnert and Andrews, 1977), and the finding was confirmed in subsequent studies (Xiao et al., 1985; Sabah et al., 1986). The time course of drug sensitivity is peculiarly biphasic: treatment of mice between days -1 and +1 of infection is highly effective, while sensitivity is already sharply diminished at week 1 after infection and reaches a minimum around week 4, when very few parasites are eliminated. Then sensitivity rises again, to reach its maximum around week 7 or later. More recent studies have re-examined the issue in terms of ED$_{50}$s, showing that 28-day-old schistosomes have a praziquantel sensitivity that is about 30 times lower than adult worms, both in vivo and in vitro (Pica-Mattoccia and Cioli, in press).

Such a pronounced age dependence of praziquantel susceptibility has an immediate impact on the efficacy of praziquantel chemotherapy in humans, as we are going to discuss next.

3.5   The question of drug resistance

Previous experience with other drugs suggests that the emergence of praziquantel resistance is probably inevitable. Anti-microbial and anti-malarial drugs provide plenty of examples to teach us that the very mechanisms of mutation and selection at the basis of biological evolution are bound to function in any population of organisms exposed to new environmental challenges. The only question seems to be the time that will elapse before schistosomes insensitive to praziquantel will outnumber the sensitive ones to the point of rendering the drug practically useless. Several factors suggest that this time may be very long, since schistosomes have a long generation time, live very dispersed, reproduce sexually and are rarely exposed to high drug selection pressure. But it is dangerous to rely on this kind of predictions. In practice, a rational approach to the threat of resistance to praziquantel should be twofold: 1- a close monitoring of the situation with the implementation of measures that may delay the emergence of resistance; 2- the timely development of alternative drugs. As we shall see, only the first approach appears to be presently pursued.

The first serious alarm that praziquantel resistance might have emerged came in the early 1990s from a very intense focus of *S. mansoni* infection in northern Senegal. Cure rates in that area were as low as 18%, as opposed to more commonly encountered rates of 80-90% (Gryseels *et al.*, 1994; Stelma *et al.*, 1995). A careful analysis of the situation, however, led to an interpretation of the data as most probably due to factors other than drug resistance. In a population exposed to very frequent transmission of the infection, drug treatment will inevitably include many individuals that have been infected recently and are thus harboring immature parasites. Since these parasites are largely insensitive to praziquantel, they will survive and contribute to the low percentage of cured individuals. This interpretation was supported by the high cure rates obtained in a subpopulation of infected people from the same focus that had moved to a transmission-free area, and by the high cure rates obtained after a second drug dose that presumably eliminated the immature schistosomes that had in the meantime grown to maturity and to drug sensitivity (Gryseels *et al.*, 2002). The hypothesis of drug resistance, however, could not be completely eliminated, since infected snails collected in that area of Senegal and brought to the laboratory gave origin to a schistosome isolate that proved to have a diminished sensitivity to praziquantel (Fallon *et al.*, 1995; Doenhoff *et al.*, 2002).

A systematic study was carried out in Egypt, where schistosome isolates were established from the eggs excreted by patients that had received three unsuccessful praziquantel doses. In many cases, these isolates had a praziquantel ED_{50} (determined in the mouse) that was 2-5 times higher than that found in isolates derived from successfully treated patients (Ismail *et al.*, 1996, 1999).

Finally, schistosomes maintained in the laboratory under continuous drug pressure, proved to acquire a decreased sensitivity to praziquantel when compared to their unselected counterparts (Fallon & Doenhoff, 1994).

Most of the schistosome isolates derived from the previously mentioned settings, plus some new ones, have been recently re-examined in a project based on the determination of the ED_{50} of each isolate in three different laboratories (manuscript in preparation). The evidence shows that isolates of schistosomes that have been exposed to praziquantel either in the field or in the laboratory have indeed a decreased drug sensitivity in comparison with isolates that have never been exposed to the drug either because they were established before the advent of praziquantel or because they came from patients that were later successfully treated. The ED_{50} differences are relatively small (2- 3-fold), but generally reproducible. No "super-resistant" schistosomes have been encountered so far and continuous drug pressure does not seem to produce further increases in the ED_{50}. It is unlikely that such a level of drug refractoriness may constitute, at the present time, a real problem for human chemotherapy, especially since the dose employed in clinical practice is –at least theoretically– close to an ED_{90} and at that dose the large majority of parasites are indeed eliminated. Close monitoring, however, seems necessary since we could be witnessing only the first step of an additive escalation to resistance (Cioli, 2000).

3.6   Drug effects

The molecular mechanism of action of praziquantel has not been elucidated, although some of the events that occur in the schistosome upon contact with the drug have been described in some detail.

The most significant phenomenon consists in a massive increase of intracellular calcium in the parasite (Pax *et al.*, 1978). Praziquantel *per se* does not have the properties of an ionophore (Pax *et al.*, 1978), and the ATPases involved in pumping Ca^{++} out of cells are apparently unaffected by praziquantel (Nechay et al., 1980). An alteration of voltage-gated calcium channels was not supported by the findings of Fetterer et al., 1980 since these authors found that Ca++ influx induced by praziquantel was not inhibited by D-600, a known blocker of calcium channel function. Even though the reasons for calcium influx are not clear, the alteration in calcium homeostasis could be the basis for two additional phenomena, i.e. muscular contraction and tegumental disruption.

A few seconds after exposure to micromolar concentrations of praziquantel, schistosomes exhibit a dramatic spastic paralysis of the musculature that is dependent on the presence of calcium in the medium (Pax *et al.*, 1978). A practical consequence of spastic paralysis is that schistosome suckers cease functioning and the parasite loses its hold on the inner wall of the blood vessel where it resides, with a subsequent liver shift (for *S. mansoni* and *S. japonicum*) or lung shift (for *S. haematobium*).

Another prominent and early event after praziquantel exposure consists in critical morphological alterations to the worm tegument (initial basal vacuolization, followed by surface blebbing and massive disruption) (Becker *et al.*, 1980). These events are largely abolished by preincubation in calcium-free medium and are not induced by the inactive praziquantel stereoisomer (Bricker *et al.*, 1983). As previously mentioned, tegumental disruption leads to exposure of new antigens and the immuno-potentiation of praziquantel.

3.7   Mechanism of action

The possibility has been considered that praziquantel may interact directly with membrane phospholipids and alter the permeability of bilayer structures. Experimental evidence for such interactions has been obtained with model membranes (Harder *et al.*, 1988; Schepers *et al.*, 1988), but it is not easy to imagine specific structures for trematode and cestode membranes that would restrict to them the activity of praziquantel. In addition, the stereo-selective activity of the drug is better reconciled with the existence of some protein receptor, rather than with a general membrane interaction.

The role of praziquantel receptor has been attributed to a popular schistosome enzyme, glutathione S-transferase, on the basis of the fact that the three-dimensional structure of this protein presents a pocket that could theoretically fit a praziquantel molecule (McTigue *et al.*, 1995). This hypothesis has been later disproved, mainly because praziquantel fails to actually modify the enzymatic activity of glutathione S-transferase (Milhon *et al*, 1997).

A recent hypothesis that is more firmly grounded on the events associated with praziquantel activity, proposes that the beta subunits of schistosome voltage-operated calcium channels are involved in praziquantel activity. This is based on structural considerations, since schistosome beta subunits have been found to possess a peculiar sequence that sets them apart from the corresponding mammalian subunits. In addition, there is functional evidence showing that schistosome beta subunits, when expressed together with alpha subunits of a different species, confer to the latter a normally nonexistent sensitivity to praziquantel (Kohn *et al.*, 2001). In an attempt to obtain confirmation for this hypothesis, the calcium channel beta subunits of schistosomes showing different sensitivity to praziquantel have been recently sequenced, but no meaningful differences could be detected between the various isolates. In addition, northern blots failed to show quantitative differences in the expression of beta subunits between schistosome isolates of different praziquantel sensitivity or between mature and immature parasites (Valle *et al.*, 2003). This negative evidence does not disprove the hypothesis that beta subunits of calcium channels may be involved in praziquantel activity, since drug insusceptibility can arise from mechanisms other than target modification, like drug uptake and drug efflux.

4. CONCLUSIONS

Only one drug, praziquantel, is commercially available today against *S. japonicum* and *S. haematobium*. Should oxamniquine be discontinued, praziquantel would remain the only drug even against *S. mansoni*. This is clearly a dangerous situation, since some evidence of praziquantel insusceptibility already exists and a full-blown resistance could appear at any moment. The need for new antischistosomal drugs is quite evident, and it is an urgent need since the discovery and the development of new drugs is a process of unpredictable length, but it is certainly measurable in several years if not in decades.

The last antischistosomal drugs were introduced in the 1970s and no systematic attempt has been made since to discover new ones (with the exception of an extremely modest investment recently made by TDR to promote compound screening against schistosomiasis). Pharmaceutical companies are not motivated to invest in research for antischistosomal drugs, since the prospects of economic returns are far from realistic. Thus, it is essential that public institutions, international organizations and charitable foundations take the lead at least in the initial stages of discovery.

The other essential requirement is that bright young minds apply themselves to unravel the fascinating enigmas that are still surrounding the schistosomes and the disease they cause in so many people.

ACKNOWLEDGEMENTS

The authors are currently receiving financial support from the INCO-II Programme of the European Commission (contract ICA4-CT-2001-10079).

REFERENCES

Allen, H.E., Crompton, D.W., de Silva, N., LoVerde, P.T., and Olds, G.R., 2002, New policies for using anthelmintics in high risk groups, *Trends Parasitol.* **18:**381-382.

Archer, S., Pica-Mattoccia, L., Cioli, D., Seyed-Mozaffari, A., and Zayed, A.H., 1988, Preparation and antischistosomal and antitumor activity of hycanthone and some of its congeners. Evidence for the mode of action of hycanthone, *J. Med. Chem.* **31:**254-260.

Becker, B., Mehlorn, H., Andrews, P., Thomas, H., and Eckert, J., 1980, Light and electron microscopic studies on the effect of praziquantel on *Schistosoma mansoni, Dicrocoelium dendriticum,* and *Fasciola hepatica* (Trematoda) in vitro, *Z. Parasitenk* **63:**113-118.

Bricker, C.S., Depenbusch, J.W., Bennett, J.L., and Thompson, D.P., 1983, The relationship between tegumental disruption and muscle contraction in *Schistosoma mansoni* exposed to various compounds, *Z. Parasitenk* **69:**61-71.

Brindley, P.J., and Sher, A. 1987, The chemotherapeutic effect of praziquantel against *Schistosoma mansoni* is dependent on host antibody response, *J. Immunol.* **139:**215-220.

Cioli, D., 1998, Chemotherapy of schistosomiasis: An update, *Parasitol. Today* **14:**418-422.

Cioli, D., 2000, Praziquantel: is there real resistance and are there alternatives?, *Curr. Op. Inf. Dis.* **13:**659-663.

Cioli, D., and Knopf, P.M., 1998, A study of the mode of action of hycanthone against *Schistosoma mansoni* in vivo and in vitro, *Am. J. Trop. Med. Hyg.* **29:**220-226.

Cioli, D., Pica-Mattoccia, L., and Archer, S., 1995, Antischistosomal drugs: past, present ... and future? *Pharmacol. Ther.* **68:**35-85.

Cioli, D., Pica-Mattoccia, L., and Moroni, R., 1992, *Schistosoma mansoni:* Hycanthone/ oxamniquine resistance is controlled by a single autosomal recessive gene. *Exp. Parasitol.* **75:**425-432.

Cioli, D., Pica-Mattoccia, L., Rosenberg, S., and Archer, S., 1985, Evidence for the mode of antischistosomal action of hycanthone, *Life Sci.* **37:**161-167.

Doenhoff, M.J., Sabah, A.A., Fletcher, C., Webbe, G., and Bain, J., 1987, Evidence for an immune-dependent action of praziquantel on *Schistosoma mansoni* in mice, *Trans. R. Soc. Trop. Med. Hyg.* **81:**947-951.

Doenhoff, M.J., Kusel, J.R., Coles, G.C., and Cioli, D., 2002, Resistance of *Schistosoma mansoni* to praziquantel: is there a problem? *Trans. R. Soc. Trop. Med. Parasitol.* **96:**465-469.

Fallon, P.G., and Doenhoff, M.J., 1994, Drug-resistant schistosomiasis: Resistance to praziquantel and oxamniquine induced in *Schistosoma mansoni* in mice is drug specific. *Am. J. Trop. Med. Hyg.* **51:**83-88.

Fallon, P.G., Sturrock, R.F., Capron, A., Niang, M., and Doenhoff, M.J., 1995, Short report: Diminished susceptibility to praziquantel in a Senegal isolate of *Schistosoma mansoni, Am. J. Trop. Med. Hyg.* **53:**61-62.

Fetterer, R.H., Pax, R.A., and Bennett, J.L., 1980, Praziquantel, potassium and 2,4-dinitrophenol: analysis of their action on the musculature of *Schistosoma mansoni, Eur. J. Pharmacol.* **64:**31-38.

Frohberg, H., 1984, Results of toxicological studies on praziquantel. *Arzneim. Forsch.* **34**:1137-1144.

Gönnert, R., and Andrews, P., 1977, Praziquantel, a new broad-spectrum antischistosomal agent, *Z. Parasitenk* **52**:129-150.

Gryseels, B., Mbaye, A., De Vlas, S.J., Stelma, F.F., Guissé, F., Van Lieshout, L., Faye, D., Diop, M., Ly, A., Tchuem-Tchuente, L.A., Engels, D., and Polman, K., 2002, Are poor responses to praziquantel for the treatment of *Schistosoma mansoni* infections in Senegal due to resistance? An overview of the evidence, *Trop. Med. Intl. Health* **6**:864–873.

Gryseels, B., Stelma, F.F., Tall,a I., van Dam, G.J., Polman, K., Sow, S., Diaw, M., Sturrock, R.F., Doehring-Scwerdtfeger, E., Kardoff, R., Decam, C., Niang, M., and Deelder A.M., 1994, Epidemiology, immunology and chemotherapy of *Schistosoma mansoni* infections in a recently exposed community in Senegal, *Trop. Geogr. Med.* **46**:209-219.

Harder, A., Goossens, J., and Andrews, P., 1988, Influence of praziquantel and Ca^{2+} on the bilayer-isotropic-hexagonal transition of model membranes, *Mol. Biochem. Parasitol.* **29**:55-60.

Ismail, M., Botros, S., Metwally, A., William, S., Farghally, A., Tao, L.F., Day, T.A., and Bennett, J.L., 1999, Resistance to praziquantel: Direct evidence from *Schistosoma mansoni* isolated from Egyptian villagers, *Am. J. Trop. Med. Hyg.* **60**:932-935.

Ismail, M., Metwally, A., Farghaly, A., Bruce, J., Tao, L.F., and Bennett J.L., 1996, Characterization of isolates of *Schistosoma mansoni* from Egyptian villagers that tolerate high doses of praziquantel, *Am. J. Trop. Med. Hyg.* **55**:214-218.

Karanja, D.M., Boyer, A.E., Strand, M., Colley, D.G., Nahlen, B.L., Ouma, J.H., and Secor, W.E., 1998, Studies on schistosomiasis in western Kenya: II. Efficacy of praziquantel for treatment of schistosomiasis in persons coinfected with human immunodeficiency virus-1, *Am. J. Trop. Med. Hyg.* **59**:307-311.

Katz, N., Dias, E.P., Araujo, N., and Souza, C.P., 1973, Estudo de uma cepa humana de *Schistosoma mansoni* resistente a agentes esquistossomicidas, *Rev. Soc. Bras. Med. Trop.* **7**:381-387.

Kohn, A.B., Anderson, P.A., Roberts-Misterly, J.M., and Greenberg, R.M., 2001, Schistosome calcium channel beta subunits. Unusual modulatory effects and potential role in the action of the antischistosomal drug praziquantel. *J. Biol. Chem.* **276**:36873-36876.

Kramers, P.G.N., Gentile, J.M., Gryseels, B.J.M., Jordan, P., Katz, N., Mott, K.E., Mulvihill, J.J., Seed, J.L., and Frohberg, H., 1991, Review of the genotoxicity and carcinogenicity of antischistosomal drugs; is there a case for a study of mutation epidemiology? Report of a task group on mutagenic antischistosomals, *Mutation Res.* **257**:49-89.

McTigue, M.A., Williams, D.R., and Tainer, J.A., 1995, Crystal structures of a schistosomal drug and vaccine target: glutathione S-transferase from *Schistosoma japonica* and its complex with the leading antischistosomal drug praziquantel, *J. Mol. Biol.* **246**:21-27.

Milhon, J.L., Thiboldeaux, R.L., Glowac, K., and Tracy, J.W., 1997, *Schistosoma japonicum* GSH S-transferase Sj26 is not the molecular target of praziquantel action, *Exp. Parasitol.* **87**:268-274.

Nechay, B.R., Hillman, G.R., and Dotson, M.J., 1980, Properties and drug sensitivity of adenosine triphosphatases from *Schistosoma mansoni*, *J. Parasitol.* **66**:596-600.

Pax, R., Bennett, J.L., and Fetterer, R., 1978, A benzodiazepine derivative and praziquantel: Effects on musculature of *Schistosoma mansoni* and *Schistosoma japonicum*, *Naunyn-Schiedberg's Arch. Pharmacol.* **304**:309-315.

Pica-Mattoccia, L., and Cioli, D., 1986, Lack of correlation between schistosomicidal and anticholinergic properties of hycanthone and related drugs, *J. Parasitol.* **72**:531-539.

Pica-Mattoccia, L., Archer, S., and Cioli, D., 1992, Hycanthone resistance in schistosomes correlates with the lack of an enzymatic activity which produces the covalent binding of hycanthone to parasite macromolecules, *Mol. Biochem. Parasitol.* **55**:167-176.

Pica-Mattoccia, L., Cioli, D., and Archer, S., 1989, Binding of oxamniquine to the DNA of schistosomes, *Trans. R. Soc. Trop. Med. Hyg.* **83**:89-96.

Pica-Mattoccia L., Cioli D. Sex- and stage-related sensitivity of *Schistosoma mansoni* to *in vivo* and *in vitro* praziquantel treatment. Intl J Parasitol 2004, in press.

Pica-Mattoccia, L., Dias, L.C.S., Moroni, R., and Cioli, D., 1993, *Schistosoma mansoni*: Genetic complementation analysis shows that two independent hycanthone/oxamniquine-resistant strains are mutated in the same gene, *Exp. Parasitol.* **77**:445-449.

Pica-Mattoccia, L., Novi, A., and Cioli, D., 1997, The enzymatic basis for the lack of oxamniquine activity in *Schistosoma haematobium* infections, *Parasitol. Res.* **83**:687-689.

Rogers, S.H., and Bueding, E., 1971, Hycanthone resistance: Development in *Schistosoma mansoni*, *Science* **172**:1057-1058.

Sabah, A.A., Fletcher, C., Webbe, G., and Doenhoff, M.J., 1985, *Schistosoma mansoni*: reduced efficacy of chemotherapy in infected T-cell-deprived mice, *Exp. Parasitol.* **60**:348-354.

Sabah, A.A., Fletcher, C., Webbe, G., and Doenhoff, M.J., 1986, *Schistosoma mansoni*: chemotherapy of infections of different ages, *Exp. Parasitol.* **61**:294-303.

Sauma, S.Y., and Strand, M., 1990, Identification and characterization of glycophosphatidylinositol-linked *Schistosoma mansoni* adult worm immunogens, *Mol. Biochem. Parasitol.* **38**:199-210.

Schepers, H., Brasseur, R., Goormaghtigh, E., Duquenoy, P., and Ruysschaert, J.M., 1988, Mode of insertion of praziquantel and derivatives into lipid membranes. *Biochem. Pharmacol.* **37**:1615-1623.

Stelma, F.F., Talla, I., Sow, S., Kongs, A., Niang, M., Polman, K., Deelder, A.M., and Gryseels, B., 1995, Efficacy and side effects of praziquantel in an epidemic focus of *Schistosoma mansoni*, *Am. J. Trop. Med. Hyg.* **53**:167-170.

Sulaiman, S.M., Traoré, M., Engels, D., Hagan, P., and Cioli, D., 2001, Counterfeit praziquantel, *Lancet* **358**:666-667.

Valle, C., Troiani, A.R., Festucci, A., Pica-Mattoccia, L., Liberti, P., Wolstenholme, A., Francklow, K., Doenhoff, M.J., and Cioli, D., 2003, Sequence and level of endogenous expression of calcium channel β subunits in *Schistosoma mansoni* displaying different susceptibilities to praziquantel, *Mol. Biochem. Parasitol.* **130**:111-115.

Waring, M. J., 1970, Variation in the supercoils in closed circular DNA by binding of antibiotics and drugs: Evidence for molecular models involving intercalation, *J. Mol. Biol.* **54**:247-279.

Watabe, T., Ishizuka, T., Isobe, M., and Ozawa, N., 1982, A 7-hydroxymethyl sulfate ester as an active metabolite of 7,12-dimethylbenz[*a*]anthracene, *Science* **215**:403-405.

Wu, M.H., Wei, C.C., Xu, Z.Y., Yuan, H.C., Lian, W.N., Yang, Q.J., Chen, M., Jiang, Q.W., Wang, C.Z., and Zhang, S.J., 1991, Comparison of the therapeutic efficacy and side effects of a single dose of levo-praziquantel with mixed isomer praziquantel in 278 cases of schistosomiasis japonica, *Am. J. Trop. Med. Hyg.* **45**:345-349.

Xiao, S.H., Catto, B.A., and Webster, L.T., 1985, Effects of praziquantel on different developmental stages of *Schistosoma mansoni* in vitro and in vivo, *J. Inf. Dis.* **151**:1130-1137.

Chapter 14

PUBIC HEALTH STRATEGIES FOR SCHISTOSOMIASIS CONTROL

Dirk Engels and Lorenzo Savioli
Strategy Development and Monitoring for Parasitic Diseases and Vector Control Communicable Diseases Control, Prevention and Eradication World Health Organization, Geneva

Key words: schistosomiasis, soil-transmitted helminthiases, deworming, helminth control, parasitic diseases control, regular large-scale treatment

1.   INTRODUCTION

Schistosomiasis remains one of the most prevalent parasitic diseases in the world. It is endemic in 76 countries and territories, and continues to be a public health concern in the developing world. Because it is a chronic insidious disease, it is poorly recognized at early stages, and becomes a threat to development as the disease disables men and women during their most productive years if left untreated. It is particularly linked to agricultural and water development schemes. It is typically a disease of the poor who live in conditions that favor transmission and have no access to proper care or prevention measures. Whereas a number of countries have managed to sustain schistosomiasis control over the last two decades, most donor-funded vertical control initiatives set up in Africa during the 1980's have proven to be unsustainable. Where control has been successful, the number of people infected and at risk of infection is very small. This is the situation in most formerly endemic countries in Asia, the Middle East and the Americas. In 2001, the 54th World Health Assembly (WHA) defined the current strategy for the control of morbidity due to schistosomiasis and soil-transmitted helminthiasis. This strategy is based on the availability of good quality anthelminthic drugs at all levels of the health care system in endemic areas, and regular treatment of groups at a high risk of morbidity.[1] Regular deworming, particularly in the younger age groups, will indeed help to avoid the worst effects of infection even in the absence of improvement in safe water supply and sanitation. The WHA also requested WHO to advocate new partnerships for implementation and to provide international direction and co-ordination with the aim of stepping up worm control and reaching the minimum target set by the 54th WHA resolution (WHA 54.19) – i.e. the regular treatment of at least 75% of school age children at risk of morbidity by 2010. In June 2001, a broad partnership– the Partners for Parasite Control (PPC) – was created. The fact that the price of anthelminthic drugs has plummeted over the last few years (currently between US$ 0.02 and 0.30 per dose) and that deworming can easily be incorporated into regular field activities, in both the education and health sectors, at little extra cost, has made the 2010 goal feasible.[2, 3]

2.   SCHISTOSOMIASIS IN THE GLOBAL PUBLIC HEALTH CONTEXT

While the distribution of schistosomiasis has changed over the last fifty years due to successful control programs, the number of people estimated to be infected or at risk of infection has not been reduced. In sub-Saharan Africa where there have been mostly short lived attempts at control and the population has increased by approximately 70% over the last 25 years, a greater number of people are infected or at risk of infection. It is estimated that about 652 million people are at risk of infection from the five human schistosome species and that 193 million are infected. Based on these estimates, 85% of the number of infected people appears to be on the African continent.[4]

There is also a general consensus today that the consequences of schistosomiasis and its global burden have been highly underestimated. A recent review study aimed at revising the global burden of diseases (GBD) due to schistosomiasis reported a mortality of over 200,000 deaths per year in sub-Saharan Africa alone.[5] For schistosomiasis haematobia the number of individuals with hydronephrosis was estimated to be close to 20 millions, and 70 millions (mainly school-age children) are thought to suffer from haematuria. For schistosomiasis mansoni the estimates suggested 8.5 million with serious liver pathology and over one million with irreversible portal hypertension.

More accurate estimates of schistosomiasis burden would therefore be highly desirable, as the GBD is being increasingly used for policy decisions regarding the allocation of resources. Both mortality and morbidity burden should be taken into account, although morbidity is expected to represent the largest share of the schistosomiasis burden. Schistosomiasis certainly represents a major cause of burden in some geographic areas. However, the GBD methodology does not provide ways of dealing with the focal nature of schistosomiasis. Better epidemiological data are needed in all endemic countries, and may have the potential to better model the heterogeneity of the schistosomiasis burden (Catherine Michaud, personal communication).

3. A GLOBAL STRATEGY FOR MORBIDITY CONTROL: BUILDING ON THE SUCCESSES AND FAILURES OF THE PAST

The 1984 WHO Expert Committee on the Control of Schistosomiasis introduced a strategy for morbidity control, which had become feasible because of the availability of effective and safe single dose drugs.[6] This created high hopes for success, not only in terms of a reduction in the burden of this disease, but also in terms of possible elimination of the infection through a presumed effect of regular treatment on transmission.

3.1 The successes

A number of countries have appreciated the public health importance of schistosomiasis and have initiated control. Some have done this long before modern single dose drugs became available, others have seized this latter opportunity to start control during the 1980's.

In China and Japan, the high morbidity and mortality due to *S. japonicum* leading to the disintegration of communities and consequent reduction in agricultural production justified control.[7,8] In Brazil schistosomiasis was one of the leading public health problems.[9] Control was initiated in Egypt because irrigation is the mainstay of agriculture and it was felt that morbidity due to schistosomiasis would hamper production.[10] In Morocco the intensive development of irrigated agriculture and the associated threat of an expansion of the schistosomiasis problem were the incentive to initiate national control.[11]

Some of these endemic countries, such as Brazil, China, the Philippines and Egypt, have been able to sustain control programs for a prolonged period and have succeeded in reducing morbidity to very low levels. Others, such as the smaller Caribbean Islands, the Islamic Republic of Iran, Mauritius, Morocco, Puerto Rico, Tunisia and Venezuela, are even nearing elimination or have already achieved this goal.[12]

3.2   The failures

During the 1980's, community-wide treatment campaigns were initiated also in Africa in numerous endemic areas. Because of the high drug prices, active diagnosis and treatment was the most cost-effective and thus the preferred strategy. After an initial "attack phase" involving substantial resources, often provided by external donors, it was expected that the endemic level would have decreased to an extent that national health authorities would be able to take over implementation during a maintenance phase. However this did not occur and donor fatigue and lack of political commitment by endemic countries replaced the early excitement for control opportunities with a feeling of frustration.

Despite the fact that the 1991 WHO Expert Committee on the Control of Schistosomiasis called for greater flexibility and a more prominent role for Primary Health Care services (PHC) and other sectors in sustainable implementation,[13] most of the schistosomiasis control activities in sub-Saharan Africa were brought to an end. Even so the lesson from these programs has been that in spite of continuous transmission morbidity can be controlled and irreversible sequelae in adulthood especially can be prevented by access to treatment and regular treatment of young individuals with reversible pathology.[2]

4.   PRESENT SITUATION, STRATEGIC APPROACH AND CHALLENGES

As the global situation today is such that schistosomiasis in many previously highly endemic areas is near to elimination and in others, mainly sub-Saharan Africa, is either expanding or has remained highly endemic, two different strategic approaches are needed.   The last WHO Expert committee has discussed and addressed this dual strategy in detail.[14]

4.1   Control in highly endemic areas

In high burden areas the main principles of schistosomiasis control are focused on morbidity control. The aspirations of the 1980s when praziquantel – the drug of choice for all forms of schistosomiasis – became available are only today materializing because of the significant drop in price. Several brands of good quality, generic praziquantel are on the market today.[15,16] The cost of an average treatment with this drug has reduced to less than US$ 0.30. This is opening up perspectives for a more generalized access to the drug in the poorest sections of the populations. As praziquantel is a safe drug,[17] it can be delivered at the most peripheral levels of the drug delivery system, also by non-medical personnel.

The current WHO strategy for the implementation of morbidity control in high burden areas[14] places emphasis on better targeting of control interventions and more cost-effective and sustainable implementation of control strategies. A simple, easy and affordable control package has been defined and adapted to the prevailing public health context in high burden countries. Integration of control in existing structures and activities, with decentralization of decision-making and delivery, are key elements in this package. Minimal implementation targets have been laid down.[1] These include access to essential anthelminthic drugs in health services in endemic areas for the treatment of clinical cases and the regular treatment of high risk group, mainly school-age children. WHO and its Member States estimate that it is feasible to provide regular treatment to at least 75% of school-age children at risk of morbidity by the year 2010.

WHA resolution 54.19 promotes an innovative public health approach in which schistosomiasis control is combined with that of soil-transmitted helminths. A recent WHO Expert Committee has recommended the practical aspects of its implementation.[14] This Committee also suggested that in specific areas other helminthic infections, such as foodborne trematode and cestode infections, be considered for control according to the same approach.

4.1.1   The rationale for the new strategic approach

Treatment with any of the anthelminthic drugs on the WHO essential drugs list (albendazole, levamisole, mebendazole, or pyrantel for soil-transmitted helminths, and praziquantel for all species of schistosomes) is safe, even when given to uninfected people. There is therefore no need to examine each individual for the presence of worms in highly endemic areas in low income countries. Individual screening offers no safety benefits and is not cost-effective in these settings - it costs four to ten times more than the anthelminthic treatment itself. It implies that presumptive treatment, on the basis of early clinical symptoms, or universal treatment on the basis of epidemiological criteria, have become cost-effective in an increasing number of endemic situations.[18,19] Early and regular treatment will avoid the worst effects of infection – subtle morbidity in children and late irreversible sequelae in adulthood - even in the presence of continued reinfection.

The provision of early clinical care is a first essential component of control within the existing health system. A symptomatic patient must indeed be able to find treatment close to his home. Today, with cost-recovery mechanisms in place, the financial burden on the health system of such treatment can be minimal. Yet, very few high burden countries have managed to provide access to praziquantel in their peripheral health infrastructure. Still, the effect of adequate case management through health services on overall morbidity is probably limited due to low health care seeking behavior of patients and low alertness of health workers about symptoms of (particularly intestinal) schistosomiasis. Other, more pro-active morbidity control efforts are therefore necessary.[20]

This type of efforts should first be targeted to school-age children. There is indeed evidence that the impact of treatment on morbidity decreases with age, and that repeated treatment in the early stages of life has an impact on subtle morbidity and a long-lasting effect on morbidity at a later age. In this perspective, focusing the delivery of regular chemotherapy on the younger age groups will produce maximum benefits, and prevent chronic sequelae in adulthood.[21-23] School-age children can be reached through the primary school system, in collaboration with the educational sector. Even in areas where school enrolment rates are low, outreach activities can be designed to ensure good coverage.[24]

School based deworming has its full impact when delivered as part of an integrated school health package such as the one promoted by the FRESH (Focus Resources on Effective School health) framework.[25] Deworming through schools is a low cost intervention. Operational research in Ghana and Tanzania has demonstrated that for the first five years of intervention, the average yearly cost of delivered treatment – taking into account current drug prices - is typically less than US$ 0.50 per child in an area where both schistosomiasis and the common intestinal worms are present, and less than US$ 0.25 per child in an area where only the latter are present. This is the total cost which includes training of teachers, as well as the procurement and distribution of drugs.[26] The high return that deworming generates in terms of health impact,[3] educational outcome[27,28] and even long-term effects on labor income,[29] fully justifies subsidizing school-based treatment.[30] The recommendations on how frequently to deliver targeted treatment to school age children in different endemic situations have recently been revised.[14]

Communities with high prevalence rates and special occupational groups such as fishermen and irrigation workers should also have easy access to regular treatment for schistosomiasis. Praziquantel can be made available to them as part of broader community based drug delivery programs or outreach services, or simply on demand through the most peripheral levels of the primary health care system (community health workers or drug distributors). The current low cost of praziquantel opens up real possibilities for cost-recovery and sustainable access.

The practical tools for epidemiological assessment and the delivery of regular treatment to high-risk groups are constantly being refined in order to meet the requirements for an easy and reliable implementation by peripheral personnel. Some, such as the questionnaire approach for rapid assessment of urinary schistosomiasis[31] or the dose pole that has been developed for praziquantel,[32] have great potential for facilitating the implementation of control in the field.

In order to enhance the effect of regular chemotherapy, long-lasting improvement in hygiene and sanitation should not be forgotten. This includes provision of safe water and sanitation, and appropriate health education. In addition, complementary integrated control activities, such as environmental management measures, should be planned with other sectors such as agriculture and water resource development programs. It is also important to ensure that any development activity likely to favor the emergence or spread of schistosomiasis and other parasitic diseases is preceded by a proper health impact assessment and accompanied by preventive measures to limit their impact.

4.1.2   Large-scale implementation by "Partners for Parasite Control"

With the new strategic approach, control of morbidity due to schistosomiasis and soil-transmitted helminths is a simple and cost-effective intervention that can be implemented through existing delivery channels. If everybody working in endemic areas were to add anthelminthic chemotherapy to ongoing activities, such as school health programs, basic curative health care and public health campaigns, morbidity could be controlled.

Over the past few years partnerships and alliances have become increasingly important in the public health field, often uniting previously disparate groups and organizations with widely different mandates and goals. There are indeed few global public health challenges where a single player has the funding, research, and delivery capabilities required to solve the problem on a worldwide scale.[33]

In resolution WHA 54.19 of 2001, WHO was requested *"to combat schistosomiasis and soil-transmitted helminth infections by advocating new partnerships with organizations of the UN system, bilateral agencies, NGOs and the private sector and by continuing to provide international direction and coordination".*[2] The *Partners for Parasite Control* (PPC) partnership was launched in June 2001 with the aim of bringing together all those with access to high risk groups to be targeted and willing to be involved in worm control, and of coordinating global control activities more effectively.[3] Active partners from day one where the World Food Program, the World Bank, the Global Parasite Control Initiative (GPCI) of Japan, CDC Atlanta (USA) and several major NGOs like the CORE group of PVOs (USA), CARE, the Partnership for Child Development, and Save the Children Fund.

The World Food Program (WFP) provides an excellent example of how deworming can be "piggy backed" on regular activities such as school feeding. WFP is progressively including deworming activities in its school feeding programs ongoing in 57 countries (www.wfp.org). The PPC received a major boost when it was joined by the Schistosomiasis Control Initiative (SCI) funded by the Bill & Melinda Gates Foundation that aims to control schistosomiasis in selected countries in Africa (www.schisto.org). Work at the World Bank is underway to integrate deworming into all FRESH-type (Focusing Resources on Effective School Health) school health programs (www.worldbank.org). The Danish Bilharziasis Laboratory (DBL) and the Japan funded Centers for International Parasite Control (CIPAC) are focusing on training and capacity building. UNICEF stepped in later focusing mainly on preschool children and women, and thus reaching key groups that are missed by the school based approach. The structure of the PPC has so far been loose, with no formal membership or governance. Nevertheless, two years after its creation, groups of core partners are emerging in different regions of the world. This is particularly visible in Africa where the above mentioned partners have started to plan and work closely together.

Controlling schistosomiasis and soil-transmitted helminthiases is not a short-term undertaking, nor is it promoted as such by the PPC. At every opportunity the partnership presses home the point that deworming must become a regular feature of a child's schooling and that control programs must be embedded in the country's infrastructure, health services and ongoing public health programs.

4.1.3   The way forward towards the 2010 goal

Despite the fact that the practical worm control package is inexpensive and can be easily implemented by the numerous partner groups and organizations in the field, the operational challenge remains enormous. It is estimated that approximately 800 million school age children are at risk of morbidity due to schistosomiasis and soil-transmitted helminth infections today. In 2002, only 3 % was covered by regular anthelminthic treatment. Taking into account the regional projections for school age populations during the years to come,[34] the number of children to be regularly treated in 2010, considering the minimal target of 75% coverage, will be close to 650 million (table 14-1).

A practical and feasible way to gain experience with deworming at the country level and to fine-tune implementation to the local conditions is by

Table 14-1. Projections (million) of the targeted school age population* in low and middle income countries

Region/country	Year 2002	2003	2004	2005	2006	2007	2008	2009	2010	Annual growth rate
Asia	323	322	322	321	320	320	319	318	318	-0.2%
of which: China	122	120	118	116	114	112	110	108	106	-1.7%
India	75	75	76	76	76	76	76	77	77	+0.3%
Latin America and Caribbean	84	85	86	87	88	88	89	90	91	+0.1%
Middle East	49	51	52	54	55	57	59	60	62	+0.3%
sub-Saharan Africa**	145	148	151	155	158	162	166	169	173	+2.3%
Total	601	606	611	617	621	627	633	637	644	

*75% of school age children at risk of morbidity[1]
**including some countries in the Middle East with a high schistosomiasis burden, such as Iraq and Yemen

developing small scale pilot initiatives that include all the components of successful implementation (epidemiological assessment, strategic planning, training at different levels and partnership building for implementation). This approach has been successful in the period 2001-2002 in 25 endemic countries that have been able - for an amount of US$ 50,000 - to cover between 50,000 and 100,000 school-age children. Once this type of practical experience is gained, it would be feasible to scale up to the national level over a period of 5-8 years provided the political commitment is built and the necessary financial effort is made.

In Guinea, the start-up phase took two years (1995-1997). After that, it took three years to reach coverage of more than 55% of the school age population. In Uganda, it took five years (1998-2003) before the first treatments started to be routinely distributed to school age children. However, it is projected that coverage of 45% will already have been reached by the end of 2004.

4.1.4   Global costs

In order to put a figure on the global financial effort needed to reach and maintain the 2010 goal, estimations were made taking into account the previously mentioned costs: US$ 0.50 per child in sub-Saharan Africa where both schistosomiasis and the common intestinal nematodes are of public health importance, and US$ 0.25 elsewhere (table 14-2). To fund further start up activities in order to cover all 105 endemic countries by 2005 would require less than US$ 5 million. The total cost to scale up to the 2010 target from 2002 onwards would be in the order of US$ 680 million. From 2010 onwards, it would require approximately US$ 200 million per year to maintain that target. Scaling up worm control according to the proposed plan would result in the delivery of over 2 billion yearly protective treatment courses by 2010 for an average cost of just over US$ 0.30 per child per year.

These estimates are by no means intended to accurately reflect what is needed at the country level. Nevertheless they are a reasonable estimate of the global cost needed to control consequences of worm infections in the developing world. They are probably even slightly over-estimated. For example in sub-Saharan Africa where over 85% of the schistosomiasis cases occur, transmission of this disease is focally distributed. Therefore not all children do need praziquantel. Peripheral health authorities in endemic countries will be able to identify areas for intervention through a simple and inexpensive epidemiological assessment method and thus save precious funding on the more expensive praziquantel.[35] For soil-transmitted helminths the maximum cost was taken into account. In some areas where only these helminths are transmitted the cost per child per year may even be lower than US$ 0.25, as illustrated by the following cost examples.

In Myanmar 200 teachers and 30 health staff were trained to distribute drugs as part of a pilot initiative. As a result of this activity over 25,000 school children received a single dose of albendazole 400 mg and health education. The total cost was US$ 1160, equivalent to less than US$ 0.05 per

Table 14-2. Cost and coverage of the proposed Global Plan to regularly deworm school-age children at risk of morbidity in low and middle income countries

	Year 2002	2003	2004	2005	2006	2007	2008	2009	2010
Start up									
Yearly number of countries/states covered	15	20	25	35					
Number of countries/states covered since 2001	25	45	70	105					
Yearly number of children covered (million)	0.75-1.5	1.0-2.0	1.25-2.5	1.75-3.5					
Yearly cost (million US$)	0.75	1.0	1.25	1.75					
Expansion									
Asia, Latin America and Middle East:									
Number of children covered (million)	12	23	46	83	130	186	257	351	471
Yearly cost (million US$)	3	6	12	21	32	46	64	88	118
sub-Saharan Africa*:									
Number of children covered (million)	4	8	15	28	44	65	91	127	173
Yearly cost (million US$)	2	4	8	14	22	32	46	64	86
Total									
Total number of children covered (million)	17	32	63	114	174	251	348	478	644
Percentage of global target reached (%)	3	5	10	18	28	40	55	75	100
Total yearly cost (million US$)	6	11	21	37	54	79	110	152	204
Total cumulative cost (million US$)	6	17	38	75	134	213	323	475	679
Number of protective treatment courses given (million)	17	49	112	226	400	651	999	1477	2121

* including some countries in the Middle East with a high schistosomiasis burden, such as Iraq and Yemen

child. This cost included the drugs, training, drug distribution and health education. The intervention is scheduled to be repeated twice a year.

In 2002-2003, the World Food Program has regularly treated over 1.7 million children in 26 countries in Africa, Asia and Latin America as part of their school feeding programs. The children were treated with mebendazole (500 mg), and praziquantel (40 mg/Kg) where necessary. The cost of including deworming in the existing school feeding programs was calculated to be between US$ 0.20 and 0.40 per child per year.

4.2   Low endemic areas

In areas where sustained schistosomiasis control efforts have resulted in significant reductions in morbidity and mortality, and a low endemic level has been reached, control efforts have to be further consolidated and schistosomiasis eliminated where possible. Where disease is no longer a public health issue, sustainable transmission control, focusing on hygiene and sanitation improvements, and environmental management, become major operational components. These will decrease the risk of resurgence of schistosomiasis and contribute to other public health goals as well.

However, such a situation also presents major challenges, as recently illustrated in China.[36] National health authorities may feel that the expenses needed to consolidate achievements are not justified anymore because of the low morbidity and therefore current public health relevance of the disease. Decentralization is often the proposed answer to this challenge in countries where control programs have been centrally run. However, peripheral authorities must be prepared and able to mobilize the resources necessary to maintain sufficient control pressure and avoid resurgence. Cost-effectiveness, decentralized decision-making and optimal use of resources by tailoring interventions to the focal transmission pattern of the disease are therefore also crucial issues in low transmission areas. As the endemic level further decreases, new objectives need to be defined, with a view to possible elimination.

Once it has been demonstrated that schistosomiasis has been eliminated from a country or an area, surveillance becomes the crucial issue. The degree of certainty that no new cases were detected also depends on the performance of the surveillance system, in terms of sensitivity of the diagnostic method used, and the operational performance of the reporting system. Surveillance should only be eased as the risk of resurgence is demonstrated to have clearly diminished or disappeared.[12]

5.   RESEARCH FOR CONTROL: SOME FURTHER NEEDS

Although tools are currently available to enable major progress in schistosomiasis control in high burden areas in the coming years, some aspects of knowledge and implementation may be further improved.

Better knowledge about the subtle and clinical disease burden, including mortality and the natural disease progression, including neglected aspects of morbidity such as genital/reproductive consequences, neurological complications and associations with other diseases, would help to better calculate the burden of disease due to schistosomiasis and raise the profile of the disease on national and international public health agendas. This is also the case for better knowledge about the economic impact of schistosomiasis and the "cost of no control".

Concerning intervention methods, improved rapid epidemiological assessment methods for intestinal schistosomiasis would be most useful. The usage and formulations of praziquantel may be subject to improvement, and the safety of its use in pregnancy will have to be further documented through pharmacovigilance.[37] As control interventions unfold, appropriate methods for post-intervention assessment and evaluation, including the effect on morbidity, will have to be defined. A new control tool to be developed might be a new schistosomicidal drug – preferably one that is active on all stages of the parasite.

Thorough documentation of control interventions in different settings will provide useful information on optimal delivery strategies and cost-effectiveness, including in areas where schistosomiasis is co-endemic with other diseases. An operational issue that will become more and more important as morbidity decreases is the use of environmental management methods in African schistosomiasis, environment-friendly focal transmission control and improved communication strategies for behavioral change.

In areas where a low endemic level has been reached as a result of sustained control interventions, the main challenge appears to be the adoption of new or improved field-applicable diagnostic methods for infection control.[12] In areas where elimination is aimed for, better knowledge about the importance of animal reservoirs, and better methods and formulations to treat infected animals, may help to speed up the process. Further improvement of environmental management strategies, and methods for surveillance in different settings and for different sub-species is also needed here.

Although expected to have a less immediate practical role in schistosomiasis control or elimination, vaccine research and development may prove to be useful in the long term.

6.   CONCLUSIONS

The global perspective presented here needs to be applied at the country level depending on the national and often sub-national needs. The current international focus given to global health priorities and the evidence such a world wide attention is built on does not reflect the very focal nature and resulting burden of certain diseases, such as schistosomiasis. The strategy currently proposed for the control of morbidity due to schistosomiasis and soil-transmitted helminthiases has been developed to enable peripheral health authorities to respond to local needs and priorities, while making optimal use of scarce resources. The global amount to be invested for control purposes is modest, given the profound benefits on health, educational outcome and personal development it will generate. It is only a minimal fraction of what is today invested in other global health priorities such as HIV-AIDS, TB and Malaria. The required funds mainly have to be mobilized at the country and/or sub-national (district) level, but in a sustainable way. We trust that the endemic countries, with the necessary support from the international community, will be willing to foot this bill.

REFERENCES

1. WHA (2001). *Schistosomiasis and soil-transmitted helminth infections.* Fifty-fourth World Health Assembly, resolution WHA54.19.
2. Savioli, L., Stansfield, S., Bundy, D.A.P., Mitchell, A., Bhatia, R., Engels, D., Montresor, A., Neira, M. and Shein, A.M. (2002). Schistosomiasis and soil-transmitted helminth infections: forging control efforts. Leading Article. *Transactions of the Royal Society of Tropical Medicine and Hygiene*, 96: 577-579.
3. The PPC Newsletter – Action against Worms. Issue 1-3. http://www.who.int/wormcontrol
4. Chitsulo, L., Engels, D., Montresor, A., Savioli, L. (2000). The global status of schistosomiasis and its control. *Acta Tropica*, 77, 41-51.
5. Van der Werf, M.J., De Vlas, S.J., Brooker, S., Looman, C.W.N., Nagelkerke, N.J.D., Habbema, J.D.F., Engels D. (2003). Quantification of clinical morbidity associated with schistosome infection in sub-Saharan Africa. *Acta Tropica*, 86, 125-139.
6. WHO (1985). The Control of Schistosomiasis. Report of a WHO Expert Committee, World Health Organization Geneva. WHO Technical Report Series No 728.
7. Chen, M.G., Zheng, F. (1999). Schistosomiasis control in China. *Parasitology International* 48, 11-19.
8. Tanaka, H., Tsuji, M. (1997). From discovery to eradication of schistosomiasis in Japan: 1847-1996. *International Journal Parasitology*, 27, 1465-1480.
9. Katz, N. (1998). Schistosomiasis control in Brazil. *Memorias do Instituto Oswaldo Cruz* 93 (Suppl. 1), 33-35.
10. Mobarak, A.B., 1982. The schistosomiasis problem in Egypt. *American.Journal of Tropical Medicine and Hygiene* 31, 87-91.

11. Laamrani, H., Mahjour, J., Madsen, H., Khallaayoune, K., Gryseels, B. (2000). *Schistosoma haematobium* in Morocco: moving from control to elimination. *Parasitology Today* 16, 257-260.

12. WHO (2001). Report of the Informal Consultation on Schistosomiasis in Low Transmission Areas: Control Strategies and Criteria for Elimination, London, 10-13 April 2000. Geneva, World Health Organization. Document WHO/CDS/CPE/SIP/2001.1.

13. WHO (1993). The Control of Schistosomiasis. Second Report of the WHO Expert Committee. Geneva, World Health Organization. WHO Technical Report Series No 830.

14. WHO (2002). The Prevention and Control of Schistosomiasis and Soil-transmitted helminthiasis. Report of a WHO Expert Committee. Geneva, World Health Organization. WHO Technical Report Series No 912.

15. Doenhoff, M., Kimani, G., Cioli, D. (2000). Praziquantel and the control of schistosomiasis. *Parasitology Today* 16, 364-366.

16. Appleton, C., Mbaye, A. (2001). Praziquantel - quality, dosages and markers of resistance. *Trends in Parasitology* 17, 356-357.

17. Dayan, A.D. (2003). Albendazole, mebendazole and praziquantel. Review of non-clinical toxicity and pharmacokinetics. *Acta Tropica,* 86: 141-159.

18. Guyatt, H., Evans, D. Lengeler, C., Tanner, M. (1994). Controlling schistosomiasis: the cost-effectiveness of alternative delivery strategies. *Health Policy and Planning* 9, 385-395.

19. Carabin, H., Guyatt, H., Engels, D. (2000). A comparative analysis of the cost-effectiveness of treatment based on parasitological and symptomatic screening for *Schistosoma mansoni* in Burundi. *Tropical Medicine and Parasitology* 5, 192-202.

20. Van der Werf, M.J. (2003). Schistosomiasis morbidity and management of cases in Africa. PhD thesis, Erasmus MC, University Medical Center Rotterdam, The Netherlands. ISBN 90-77283-02-1.

21. King, C.H., Muchiri, E.M., Ouma, J.H. (1992). Age-targeted chemotherapy for control of urinary schistosomiasis in endemic populations. *Memorias do Instituto Oswaldo Cruz* 87, 203-210.

22. Hatz, C.F., Vennervald, B.J., Nkulila, T., Vounatsu, P., Kombe, Y., Mayombana, C., Mshinda, H., Tanner, M. (1998). Evolution of *Schistosoma haematobium*-related pathology over 24 months after treatment with praziquantel among school children in southeastern Tanzania. *American Journal of Tropical Medicine and Hygiene* 59, 775-781.

23. Frenzel, K., Grigull, L., Odongo-Aginya, E., Ndugwa, C.M., Loroni-Lakwo, T., Schweigmann, U., Vester, U., Spannbrucker, N., Doehring, E., 1999. Evidence for a long-term effect of a single dose of praziquantel on Schistosoma mansoni-induced hepatosplenic lesions in northern Uganda. *American Journal of Tropical Medicine and Hygiene* 60, 927-31

24. Montresor, A., Ramsan, M., Chwaya, H.M., Ameir, H., Foum, A., Albonico, M., Gyorkos, T.W., Savioli, L., 2000. Extending anthelminthic coverage to non-enrolled school-age children using a simple and low-cost method. *Tropical Medicine and International Health* 6: 535-537.

25. The FRESH Toolkit, Focusing Resources on Effective School Health (2002). World Bank, Washington DC.

26. Partnership for Child Development (1999). The cost of large-scale school health programmes which deliver anthelmintics to children in Ghana and Tanzania. *Acta Tropica,* 73: 183-204.

27. Nokes C. and Bundy D. (1993). Compliance and absenteeism in schoolchildren: implications for helminth control. *Transactions of the Royal Society of Tropical Medicine and Hygiene,* 87:148-152.
28. Miguel E. & Kremer M. (2002) Worms: Identifying Impacts on Health and Education in the Presence of Treatment Externalities. http://post.economics.harvard.edu/faculty/kremer/
29. Bleakley, H (2002) Disease and Development: Evidence from hookworm eradication in the American South. Report of the Rockefeller Sanitary Commission. http://web.mit.edu/hoyt.
30. Economic Report of the President, transmitted to the Congress together with the Annual Report of the Council of Economic Advisers. United States Government Printing Office. Washington, February 2003.
31. Lengeler, C., Utzinger, J., Tanner, M. (2002). Screening for schistosomiasis with questionnaires. *Trends in Parasitology* 18:375-7.
32. Montresor, A., Engels, D., Chitsulo, L., Bundy, D.A.P., Brooker, S., Savioli, L. (2001). Development and validation of a "tablet pole" for the administration of praziquantel in sub-Saharan Africa. *Transactions of the Royal Society of Tropical Medicine and Hygiene* 95, 1-3.
33. Developing Successful Global Health Alliances. Bill and Melinda Gates Foundation: April 2002.
34. Interpolated population by sex and single years of age, for every calendar year 1950-2050 (2001). Supplementary tabulation to World Population Prospects: The 2000 Revision. United Nations Population Division, New York.
35. Montresor, A., Crompton D.W.T., Gyorkos T.W. & Savioli L. (2002). Helminth control in school-age children: a guide for managers of control programmes. ISBN 92 4 154556 9. World Health Organization, Geneva.
36. Jiang, Q.W., Wang, L.Y., Guo, J.G., Chen, M.G., Zhou, X.N. & Engels, D. (2002). Morbidity control in China. *Acta Tropica,* 82: 115-125.
37. WHO (2002). Report of the Informal Consultation on the use of praziquantel during pregnancy/lactation and albendazole/mebendazole in children under 24 months. Geneva, 8-9 April 2002. Geneva, World Health Organization. Document WHO/CDS/CPE/PVC/2002.4.

Chapter 15

SCHISTOSOMIASIS: ARE WE READY FOR A COORDINATED RESEARCH AND PUBLIC HEALTH AGENDA?

Daniel G. Colley
Center for Tropical and Emerging Global Diseases and Department of Microbiology, 623 Biological Sciences Building, University of Georgia, Athens, GA, 30602-2606

Key words: coordination, development plan, training, basic research, public health

1.   INTRODUCTION

As exemplified by the breadth of topics and perspectives presented in the chapters of this book, exciting research and critical public health initiatives related to schistosomiasis are actively moving forward. The authors of this volume make it abundantly clear that this is an exhilarating time to be involved in working out the intricacies of schistosome host/parasite systems and to be doing something about it. Nevertheless, in the face of this demonstration, I would like to raise a cautionary flag and then attempt to outline a potential yellow-brick road (Metro-Goldwyn-Mayer, 1939) -- down which those who care about schistosomiasis scientifically and those who work more directly to see its demise as a public health problem, might wish to travel, together.

The cautionary flag concerns the size of the aggregate community of those who work on various aspects of schistosomiasis, and it is this: there are not very many of us. Yes, this volume indicates that there are very interesting and very important aspects of schistosomiasis being investigated and acted upon, and there are many to be considered in the future. However, when viewed objectively, given the potential scope of scientific endeavors that could be tackled by those who might work on schistosomiasis, and the enormity of the challenge, it seems clear to me that the vineyard is ready for harvest, but the laborers are few. That is to say, relative to the size of the problem and to its inherent biological interest, there is a small contingent of those in biomedical research or field-based public health with an emphasis on this disease, and the number is not increasing. There are, at least in the United States and I believe around the world, fewer funded investigators working on schistosomiasis now than there were 20 years ago (and the field was not overpopulated then). Some data to back-up this claim, and some of the possible reasons for this have been discussed in regard to work on helminthic infections in general (Colley, et al., 2001). In essence the paucity of trainees who choose to enter or remain in this field seems to be related to several reasons and perceptions:

1) It is too complex to work on such a chronic, multi-host/parasite system;

2) It is too difficult to obtain sustained funding for work in a chronic, multi-host/parasite relationship (Corollary A, it is impossible to obtain or keep a position if you work on this; Corollary B, this is not HIV, tuberculosis, or malaria and will never get appropriate funding);

3) It is not possible to study schistosome development without in vitro model systems;

4) It is too difficult to ask pertinent, sophisticated questions of schistosomes without appropriate genetic tools and manipulations;

5) The impression that large-scale drug treatment programs have (or will) solve the public health problem, and take away any incentives there might have been to fund this type of research;

6) All the money goes to "pie-in-the-sky" basic research;

7) All the money goes to huge public health schemes that are never-ending and doomed to fail in the long run.

If there really is (or will be) a shortage of investigators and public health professionals in schistosomiasis, and if some of these perceptions are responsible (whether they are valid or not), then the question needs to be asked, "Can anything be done about it?" I will try to provide some background on the not so hypothetical perceptions listed above, and then propose how the greater community that cares about this infection might develop strategies to address them, and to move ahead effectively.

2. THE PERCEPTIONS AND REALITIES (AS I SEE THEM)

2.1 It is too complex to work on such a chronic, multi-host/parasite system

There is no question that studying schistosomiasis is a scientific challenge. This is because of its complex, metazoan, large genome-containing parasite, its longevity within its hosts, its multiple life-cycle forms, and its snail and mammalian hosts. There is no question that doing immunology on a chronic disease that takes many weeks to manifest some of its most intriguing aspects requires time, patience and ingenuity, and the same is true of applying other disciplines to schistosomiasis. However, from the perspective of interesting biological questions, all these aspects provide a vast number of inherently fascinating questions – all centered on "How on earth does it do that?" Therefore, in this case, I think the perception is true. It <u>is</u> complex. However, the thrill of trying to unravel and understand a fabric that has taken a very long evolutionary time to weave necessitates putting up with the vicissitudes of this intriguing system. Some help with some aspects of these challenges will be forthcoming in the future (see discussions of perceptions #3 and #4).

2.2 It is too difficult to obtain sustained funding for work in a chronic, multi-host/parasite relationship (Corollary A, it is impossible to obtain or keep a position if you work on this; Corollary B, this is not HIV, tuberculosis, or malaria and will never get appropriate funding)

I believe it is only realistic to expect that the availability of funding for scientific studies on schistosomiasis will be dependent on the quality of the scientific questions to be asked, rather than on the magnitude of the problem to be potentially addressed. But I do think we can at least expect to get good science that is related to schistosomes funded. Translated, this means that investigators who will get positions and funding to work on schistosomiasis will need to be excellent cell biologists, biochemists, molecular biologists, bioinformatists, immunologists, etc., rather than schistosomologists. This trend has long been developing and is not going to go away. [This "reality" does not address the ancillary challenge of who will teach these future scientists their parasitology, but that is a different, albeit related, fight, and will have to wait for another day.] The trend away from parasitology as a discipline can be bemoaned, but in most settings is already a fact of life. Few departments remain that retain the name Parasitology, but scattered through departments of microbiology, immunology, genetics, biochemistry, molecular biology, cell biology, geographic medicine, infectious diseases and the like are very accomplished investigators who have chosen as their model system schistosomiasis, and who apply the tools of their trade effectively to its challenges. It is based on their contributions to biomedical science, using this system, that they obtain positions and funding. Work on schistosomiasis has gotten, and does get, funded. It has, at various times, enjoyed focal attention by various foundations and philanthropies (the Edna McConnell Clark Foundation, the Rockefeller Foundation, the MacArthur Foundation) which is much appreciated, and individual projects and programs on schistosomiasis have also had sustained funding from the NIH, WHO/TDR, the Wellcome Trust, the Danish Bilharzia Laboratory, CNRS, the European Union, the Burroughs Wellcome Fund, and other programs. However, schistosomiasis is not HIV, tuberculosis or malaria. While the 200 million people who have it should not have to have it, and do deserve attention, it is not a frightening killer that startles Ministries of Health and it is never likely to garner the popular level of support that some other diseases should – based purely on its imminence as a public health threat. Still, it is a widespread, major endemic public health problem and challenge, and a fascinating scientific enigma. It should and can develop an appropriate funding base in its own right.

2.3   It is not possible to study schistosome development without in vitro model systems

It is possible, but difficult, to study a metazoan worm that is not free-living, but instead involves disparate hosts and has evolved intricate mechanisms to deal with (and capitalize on) the multiple different, potentially hostile environments in which it must develop, live and procreate. Still, some progress is being made toward in vitro culture of some stages of schistosomes, and it is strongly recommended that this be a major future goal in which investments are made (see below).

2.4   It is too difficult to ask pertinent sophisticated questions of schistosomes without appropriate genetic tools and manipulations

There is no question that it is difficult to ask many of the questions that beg to be asked of a system that has had little success, as yet, in regard to gene manipulations. Other systems have moved ahead based on the abilities to accomplish transformations, transfections, interference RNA, gene knockouts, gene knock-ins, sophisticated mRNA studies with multiple microarrays, SAGE and the like, complex functional genomics, and proteomics. The good news is that even given the inherent challenges of the schistosome systems some of these things are now beginning to be done successfully, as exemplified in Chapters 2, 3, 4 and 6 of this volume. Furthermore, as more genomic information is acquired on schistosomes, and if more intellectually stimulated young scientists for whom these are standard technologies are recruited to schistosomiasis, we can fully expect this trend to continue – and to pull the rest of the field along with it. A real emphasis needs to be made on developing and advancing these technologies so they can be applied to schistosomes and schistosomiasis.

2.5   The impression that large-scale drug treatment programs have (or will) solve the public health problem, and take away any incentives there might have been to fund this type of research

As pointed out in several chapters, the current goal for the public health community, as stated in World Health Assembly resolution 54.19 (see Chapter 14) is to control morbidity, especially in high-risk groups, through regular treatment of those at risk with anti-schistosomal drugs (which currently means praziquantel). This is a good policy. It works and in many settings is feasible. Efforts such at the Bill and Melinda Gates Foundation-funded Schistosomiasis Control Initiative seeks to demonstrate both the feasibility and effectiveness of such programs at the countrywide level in several sub-Saharan African countries. Also, working together effectively through efforts such as the Partners for Parasite Control (see Chapter 14) can help bring these desires to fruition. It is clear that we currently have a safe, effective and relatively inexpensive drug that can, when administered regularly (annually or bi-annually?) reasonably keep morbidity at bay for most of those infected and re-infected. This is the first priority of public health, to keep people from developing serious consequences due to a given disease or problem. It does not, however, address subsequent possible goals, such as elimination or eradication, by stopping transmission. Having a good tool is not the same as having the ultimate tool (or tools). Having a good tool does not even mean that there could not be better use of that tool. As is emphasized in Chapter 13, we should also not be so naïve or complacent as to think that the tools we currently use so effectively will always be efficacious or available.

Even in the most successful public health eradication campaigns (such as those against smallpox, guinea worm, and polio) it has always been true that continued research was needed in support of mid-way corrections, and adjustments to include the implementation of new and better tools (Fenner, et al., 1998; Henderson, 1998). It is, unfortunately, true that the eradication (or even approaching eradication) of a disease, or the solving of a problem is often construed as a reason to stop funding basic research on that topic. I say this is unfortunate, because we always seem to learn later that such a decision was a mistake (Opinion, 2001; Clarke, 2001). This is because unforeseen situations almost inevitably arise (be they biological, sociological or environmental) that thrust the vanquished foe back in the limelight and sometimes onto center stage. Even if the public health/medical reasoning for funding schistosomiasis were to disappear (which is not very likely in the foreseeable future, due to the realities of poverty and sanitation), studies on this fascinating parasite and its interactions with its hosts should continue to be funded because of what they can tell us that could relate to other existing biological systems (be they other infections, autoimmunity, tumors, transplants, parasitism in general, evolution, phylogeny, etc.). The funding base may shift, but should still continue, if compelling scientific questions are still being asked using this system.

2.6   All the money goes to "pie-in-the-sky" basic research

The lament of some public health workers has at times been that the search for a magic bullet (such as a vaccine or a new drug) is more expensive and less likely to yield something positive than putting the available funding on the implementation of existing tools, to the immediate benefit of those currently suffering. There is no question that vaccine or drug discovery and development are expensive, time consuming endeavors. There is also no question of the truth, thus far, of that oft-repeated adage (originally stated by Dr. Franz von Lichtenberg) that immunology has gotten more from schistosomiasis than schistosomiasis has gotten from immunology. We clearly now know and are learning more every day from studies on schistosomiasis (see Chapters 5, 7-11) about Th2-ness and its immunoregualtion, and how they might apply to basic immunology and other immunologic problems (atopy, autoimmunity) than we would have without studies on schistosomes. I maintain that this is good, not bad, and refer the reader back to perspective # 2 above. It is, of course, also hoped that such studies will eventually benefit those with, or at risk of, schistosomiasis, and efforts to translate findings into realities need to be encouraged and pursued. However, to date it must be said that huge amounts of funding (compared to what industry puts into their priority discovery and development programs) have not yet been spent on schistosome vaccine or drug discovery and development. This is to say, the amount of money that has been thus far "lost" to either basic or applied research from potential implementation programs is miniscule compared to the cost required for actual implementation of a District-, no less a Country- or Continent-based campaign, and in reality the one endeavor is not a threat to the other, nor should it be seen as such (see below).

2.7   All the money goes to huge public health schemes that are never-ending and doomed to fail in the long run

Until the recent advent of the SCI, researchers could find little truth in this perspective, in regard to the amounts spent by governments, Non-governmental organizations (NGOs), or philanthropies on the control of schistosomiasis. It is true that Brazil, Egypt and China, as well as some other countries (often with international funding assistance) have mounted major campaigns against schistosomiasis. However, none of these efforts took funds directly out of research funding pools, but rather out of development accounts that stood little or no chance of ever being spent on any type of basic research. In some instances, far thinking developmental planners did use some of these funds for at least translational research, in hopes of keeping new findings in the pipeline toward implementation. Although pervasive, perspectives #6 and # 7 are nothing more than reciprocal faces of the same mistrust/paranoia scenario. They are myths that are based on misunderstanding and mistrust. They can be, and need to be, effectively dealt with jointly by these two essential, but sometimes estranged, communities.

3. WHY WOULD WE WANT TO FORGE A COORDINATED WAY FORWARD FOR SCHISTOSOMIASIS RESEARCH AND CONTROL ?

The spectrum that is sometimes described between research and control (Colley, 1993) is, unfortunately, often more polarized into camps than a true continuum. The reasons for this polarization range from honest differences of opinion, to flagrant mistrust of one another's motives. In short, research scientists and public health practitioners are people, and people are always the problem – and the solution. Resource limitations unnecessarily pit control versus research efforts, and even foster unhealthy competition within various factions within the research community (e.g. drugs vs. vaccines; laboratory vs. field research) and the control community (e.g. school-based vs. community-based; sanitation vs. drugs; integrated vs. stand-alone programs). The schistosomiasis research and public health communities have never dealt with the potential seeds of their disharmony, nor (in fact) is it always as evident as I am portraying it. Nevertheless, we do live in a world of resource limitations, and I propose that one way to try to focus on improving the resource situation for work on schistosomiasis is through coordination and a jointly determined set of strategies to deal with the opportunities of today and the needs for tomorrow. I am proposing that it is time to bring the entire spectrum of the schistosomiasis community together, into an effort to: 1) develop an integrated global research and control strategy for schistosomiasis; and, 2) influence a variety of donors and agencies that because we have a sound plan for the future, they should assist in its implementation.

4. WHAT MIGHT GO INTO SUCH A PLAN?

I think (but others might well disagree and/or have totally different ideas) that such an overall strategy should address some of the issues raised in perceptions #1-#7 above, and outline how they might be dealt with in the future. Some of these are scientific issues, such as the need for sustained efforts to develop better animal models, appropriate cell lines and culture methods to replicate the local developmental environments of schistosome life-cycle stages, and to develop new genetic tools and manipulations for use on schistosomes. Other issues would be more translational in nature, such as support for Phase II and Phase III trials of either potential vaccines or drugs. Still other aspects that need to be included are on the implementation end of the continuum, and might included issues such as training public health supervisors and field workers, and questions in regard to the practicalities and effectiveness of school- or community-based drug distribution in different settings. Considerable effort will be required to encompass major needs and issues in a way that would garner broad-based support for such a collaboratively developed plan. It is fundamental to the success of any such enterprise that it has very broad-based support. Without it, the plan would be useless and it would rightfully fail to impress those to whom it was proposed. Dissention is not an attractive thing to fund, except, apparently, for the likes of "reality" television, which appears to be a disgusting, artificially polarizing activity that is neither satisfying nor effective in reaching a goal. In short, this is a model to be avoided.

5. THE PROPOSAL

I suggest that representatives from across the breadth of the schistosomiasis community propose a series of meetings to develop a coordinated research to control plan. These meetings would begin with a manageable-sized group of 10-12 scientists and public health officials, plus a few representatives of the most critical agencies. Its purpose would be to draft an outline of a global plan for schistosomiasis research and control, and to be as comprehensive as practical. This would be followed by a series of meetings to vet the plan with focus groups representing various key aspects (e.g. drug development, basic biology, vaccine development, operational research, control programs). These meetings would then further develop and refine the aims within the plan. A consensus plan would then be put to the wider community for comments and suggestions. This might be done either at another meeting or electronically. After feedback, the coordinated plan would then be submitted to a range of potential agencies to garner support for its various parts. The exercise of collaboratively pulling together an overarching plan that would stretch from basic research, through applied and translational research, to actual implementation of control programs, will require the building of a level of trust that in itself should be worthwhile. The ability to go to various funding agencies with such a coordinated plan and ask for support for the portions relevant to each agency should build their confidence that these are well thought out goals, worthy of their support. A plan should not be a straightjacket. The purpose of this process would be to develop a thoughtful, coherent avenue of approach to the challenge of schistosomiasis that would be coordinated to a level that would enlist the efforts of those who can and want to contribute to its goals (both participants and donors). Such a plan would always be open to innovations and redirections, because as it was decided upon, it could be re-decided upon when new input dictated the desirability to do so. It should be designed to be a "yellow brick road" (Metro-Goldwyn-Mayer, 1939) along which we can discover ourselves and by which we can attain the scientific and public health goals we establish.

REFERENCES

Colley, D.G., "Targeted research:" An oxymoron that needs to be discussed. Amer. J. Trop. Med. Hyg., 50:1-12, 1993.

Colley, D.G., LoVerde, P.T., and Savioli, L., Medical helminthology in the 21st century. Science, 293:1437-1438, 2001.

Clarke, T, Polio's last stand. Nature, 409:278-280, 2001.

Fenner, F., Hall, A.J., Dowdle, W.R., What is eradication? in *The Eradication of Infectious Diseases*, W.R. Dowdle and E.R. Hopkins, Eds (John Wiley & Sons, Chichester), pp. 3-17, 1998.

Henderson, D.A., Eradication: lesions from the past., Bull. WHO 76 (Suppl.2): 17-21, 1998.

Metro-Goldwyn-Mayer, *The Wizard of Oz*, 1939.

Opinion, Don't underestimate the enemy. Nature, 409:269, 2001.

Index